FLOODS OF FORTUNE

FLOODS OF FORTUNE

ECOLOGY AND ECONOMY ALONG THE AMAZON

MICHAEL GOULDING

NIGEL J. H. SMITH

DENNIS J. MAHAR

COLUMBIA UNIVERSITY PRESS • NEW YORK

COLUMBIA UNIVERSITY PRESS
New York Chichester, West Sussex

Map by Edanart on page 5 from *Amazon: The Flooded Forest* by Michael Goulding (London: BBC Books, 1989).

Map on page 10 by Joel LeMonnier from "Flooding Forest of the Amazon" by Michael Goulding, *Scientific American* (March 1993). Copyright © 1993 by Scientific American, Inc. All rights reserved.

Library of Congress Cataloging-in-Publication Data

Goulding, Michael.
Floods of fortune : ecology and economy along the Amazon / Michael Goulding, Nigel J. H. Smith, Dennis J. Mahar.
p. cm.
Includes bibliographical references (p.) and index.
ISBN 0-231-10420-0 (cloth).
1. Endangered ecosystems—Amazon River Valley. 2. Man—Influence on nature—Amazon River Valley. 3. Floodplain ecology—Amazon River Valley. 4. Amazon River Valley—Environmental conditions. 5. Ecosystem management—Amazon River Valley.
I. Smith, Nigel J. H. II. Mahar, Dennis J. III. Title.
QH77.A53G68 1995
574.5'0981'1—dc20 95-10904

Casebound editions of
Columbia University Press books
are printed on permanent
and durable acid-free paper.

Printed in Hong Kong
c 10 9 8 7 6 5 4 3 2 1

CONTENTS

ACKNOWLEDGMENTS

We wish to express our thanks to the Technical Department for the Latin American and the Caribbean Region of the World Bank, the Rainforest Alliance, the W. Alton Jones Foundation, and the Tinker Foundation for institutional and financial support; to Márcio Ayres, Thomas E. Lovejoy, Anthony Anderson, and Richard Evans Schultes for critical and useful comments on the manuscript; to Ed Lugenbeel, Laura Wood, Anne McCoy, and Hal Dalby at Columbia University Press for their special interest in this book; to freelance editor Connie Barlow for diving deep into the subject matter and lending her expert assistance to improve the manuscript; and to Scientific American and the BBC for allowing the use of illustrations previously published.

Michael Goulding thanks Dan Katz, Guilherme de La Penha, Pete Myers, Renate Rennie, Efrem Ferreira, Ronaldo Barthem, Cristina Esposito, Mirian Leal Carvalho, Carlos Araújo-Lima, Raimundo Aragão Serrão, Claudia Sobrevila, and Marguerite Holloway for their help. Nigel Smith expresses appreciation to the Brazilian agricultural research system (EMBRAPA) in Belém and the agricultural extension service (EMATER) in Santarém for providing logistical support for field trips in the middle Amazon. Smith has benefited from discussions with several scientists at CPATU-EMBRAPA on agriculture and forest extraction along the Amazon floodplain, particularly Adilson Serrão, Italo Falesi, and Luciano Marques. In the Santarém area, Smith learned much about agriculture and landscape change along the Amazon from conversations with the mayor, Ruy Corrêa, and Pedro Marques de Azevedo of EMATER. Smith also thanks the World Bank's Environment Division for Latin America and the Caribbean for providing opportunities to visit the Amazon floodplain three times in 1993 and 1994. Dennis Mahar thanks George Martine and Bruno Pagnoccheschi of the Society, Population, and Nature Institute in Brasília for generously providing office space and the use of their excellent library. He, too, has enjoyed the assistance of Mayor Ruy Corrêa of Santarém. Mahar gratefully acknowledges the assistance of Sri-Ram Aiyer, director of the Technical Department for the Latin American and the Caribbean Region of the World Bank.

FLOODS OF FORTUNE

AN ENDANGERED TREASURE

The Amazon Basin is a landscape of mysteries. It contains the world's biggest river and most diverse ecosystems. But in its vastness it is also perhaps the least understood.

As the new century approaches, the Amazon is being transformed by deforestation, urban growth, mining, dams, and widespread exploitation of its natural resources. A growing number of environmentalists, politicians, journalists, and even the general public in many countries now watch and worry about these changes. In a sense, the infusion of the Amazon Basin into the global environmental conscience marks a significant advance for civilization, as it underscores the mounting international concern for conserving biodiversity.

For at least a decade the world has been captivated by the ever-changing interplay of economic forces in Amazonia and the frightening ecological repercussions. The mighty Amazon River and its immense floodplain, however, have been viewed only as secondary backdrops to the economic and ecological drama unfolding on the uplands. A combination of heavy deforestation in the Brazilian states of Rondônia and Acre in the 1980s, an uncontrolled gold rush in headwater areas, the disruption of Indian tribes and their lands, and large hydroelectric projects in the tributaries have captured the spotlight of international environmental concern.

Less is now known about the Amazon River than several of its tributaries. Yet the ecological disturbance that the Amazon River, and especially its floodplain, has suffered equals if not surpasses that caused by upland deforestation, gold mining, and dam building. Floodplain settlement, farming, ranching, fisheries, and logging have already had major and, in some cases devastating, effects on the ecology of much of the

The flooded forest. *Flooded forests used to cover half of the Amazon floodplain (the other half was floating meadows and floodplain lakes). Deforestation has taken a heavy toll.*

Amazon River. The Amazon River floodplain has undergone more environmental change in the last two or three decades than in all of previous human history.

Unlike most of the upland rainforest, the Amazon River floodplain has rich soils that could boost food production. Moreover, that richness is annually renewed by the sediment-carrying floods. The draw of rich, renewable soils is a mixed blessing for those who would exploit it. Seasonal inundation brings special burdens to floodplain farmers and ranchers. Nevertheless, cattle and water buffalo production on the Amazon River floodplain are dramatically on the rise. In our view, large-scale ranching is the greatest threat to the primary economic resource the floodplain has to offer: a cornucopia of fish.

Although many Amazon fishes are harvested primarily in the main river channels and far downstream in the estuary, they ultimately depend on the floodplain. Many commercially valuable species spawn only in the floodplain; predators rely on prey that migrate seasonally from the floodplains to the river channels; juveniles rely on the fractal nature of floodplain habitat and vegetation for food and cover; and many fishes gorge on fruits and seeds that fall during the floods—fasting in the main channels during the months of low water. Of all economic activities, deforestation for pasture, along with the trampling of floodplain meadows by livestock, poses the greatest threat to established and potential fisheries. Advocates of biodiversity protection in the Amazon floodplain should thus look for natural allies in the local peoples, urban centers, and industrial interests that depend on fisheries for income or protein.

FOCUS ON THE FLOODPLAIN

To most people the word *Amazon* conjures images of one vast rainforest. This perception is erroneous, however, because the Amazon Basin is an intricate mosaic of interconnected ecosystems, including some altogether unique rainforests. But even scientists do not fully understand the intricacy. Technical knowledge of the geology and geography of the Amazon Basin, and the distribution of its plants and animals, is still too poor to identify all the ecosystems contained in the watershed. And even more worrisome, calculations of the annual rate of deforestation and the cumulative area cleared, no matter how accurate, can be used for only very crude estimates of how serious the destruction of biodiversity might be.

The destruction of biodiversity cannot be extrapolated directly from a deforestation estimate unless the various types of rainforests that have been affected are considered. To date no such analysis has been attempted. Of special concern is the degree of species uniqueness, or endemism, and the seasonal dependence of migratory animals on special types of rainforests. Computerized databases are now making it possible to detect some areas of high endemicity, but the classification of most Amazonian plant and animal groups is only at the stage prevailing in the United States and Europe in the late nineteenth century. Thus we still lack the ability to detect the destruction of endemic species for most of the Amazon Basin.

Of particular import is the fate of highly specialized rainforests, of which there are several distinct kinds in the Amazon Basin. The overall threat to biodiversity attributable to human intrusions in these specialized rainforests is much greater than that indicated by total deforestation rates for the entire basin. The status of biodiversity loss and

The most endangered habitat in the Amazon Basin.

impending damage is thus probably even more disturbing than commonly reported.

Although some of the specialized rainforests may seem small by Amazonian standards, they are still enormous. The floodplains, including their seasonally flooded forests, constitute 2 to 3 percent of the Amazon Basin, but even this fraction is substantial. These floodplains flank the main river and its tributaries over an immense landscape, stretching nearly to the Andes in the west. How much endemicity and uniqueness is found within this floodplain forest? How many kinds of specialized floodplain forest are there?

Deforestation. *The destruction of floodplain rainforests is one of the most serious ecological threats the Amazon now faces. Most of the middle region of the Amazon River floodplain has already been deforested.*

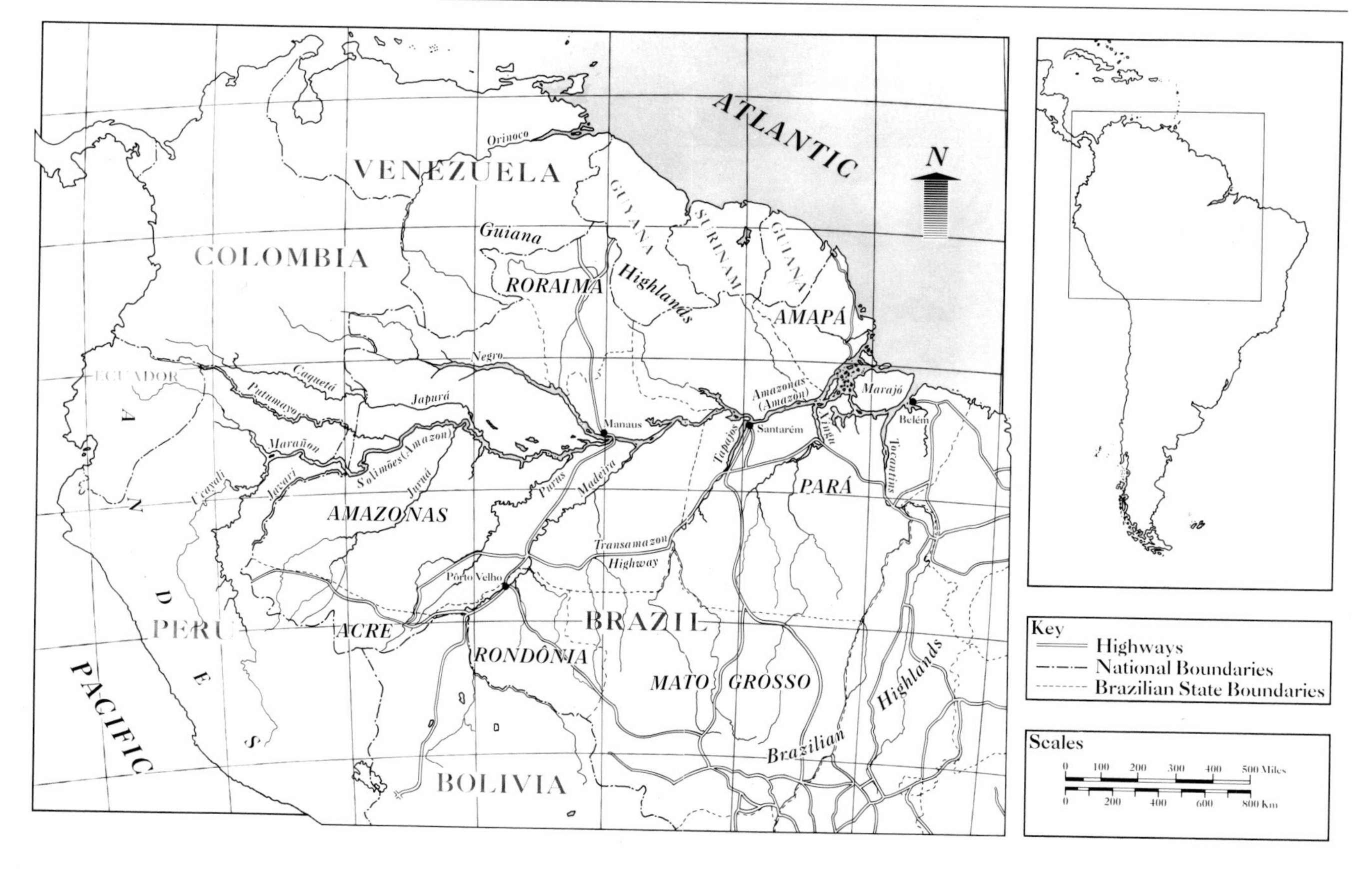

Three fundamentally different kinds of waters feed the forests of the floodplain. Rivers may be sediment-rich, clear, or darkened by tannins. These differences in water chemistry have profound effects on the abundance and distribution of life.

The Amazon itself, from the headwaters all the way to the estuary, is a sediment-rich, or muddy, river. But it is an odd color of muddiness—like coffee with cream. Because of the milky color of the Amazon and some of its tributaries, local peoples sometimes call these "whitewater" rivers—which is a confusing term for English speakers, for whom the term whitewater means roiling rapids. Henceforth we will refer to these as *muddy* or *sediment-rich rivers.*

All the great rivers of Amazonia with headwaters in the Andes are sediment-rich rivers. Among these are the Rio Juruá, Rio Purus, and Rio Madeira. For these rivers, the high, youthful mountains with their diverse geology (including volcanic soils) provide a huge sediment load and much richer nutrient base than do the sources of any of the other Amazon tributaries.

Rivers and streams that drain uplands to the east of the Andes are usually clear or almost so. These waterways arise in the old, worn-away mountains of the Brazilian Highlands to the south and the Guiana Highlands to the north. Chemically these *clearwater rivers* range from acidic to nearly alkaline, but in general they are relatively nutrient-poor. The three largest clearwater rivers are the Rio Tapajós, Rio Xingu, and Rio Tocantins. All drain the Brazilian Highlands and enter the Amazon River from the south.

The third and final category of water type is that of *blackwater rivers*, which are free

The sediment-rich Rio Madeira. *The Rio Madeira (shown here near Porto Velho) is the largest of all sediment-rich tributaries that drain the Andes.*

The clearwater Rio Tapajós. *Some clearwater tributaries are substantial, but they are quickly absorbed into the sediment-rich stream of the Amazon River.*

of heavy sediment but dark in color. They owe their special character to the millions of years of erosion that have ground down the Brazilian and Guiana highlands into sands that now fill the Amazon Basin. Because of millions of years of weathering, the sandy soils are very poor in nutrients. The streams that are born on them are among the most chemically pure on Earth. Why, then, the blackness?

It turns out that the sandy soils are too poor in microorganisms (especially fungi,

The blackwater Rio Negro. *The confluence of the Rio Negro and Amazon River marks one of the most spectacular "meeting of the waters" in the world. Being warmer (and thus lighter), the Rio Negro "rides" on top of the Amazon River until turbulence mixes it many kilometers downstream.*

bacteria, and invertebrates) to easily decompose organic matter. Organic chemicals, moreover, that would easily bind to clay sediments have no affinity for sand. The organic overload thus seeps through soils and is carried into streams and rivers, rendering some as dark as tea. Blackwater rivers are highly acidic, with a pH usually below 4.0.

Blackwater tributaries predominate in the western and central part of the Amazon Basin, while clearwater rivers are more common in the east. The largest and most famous blackwater river in Amazonia is the Rio Negro ("black river"). No mere tributary, it is actually the fourth largest river in the world. Nevertheless, within thirty kilometers of emptying into the Amazon, the black waters of the Rio Negro are rendered indistinguishable from the rest of the muddy Amazon.

Despite some overlap, each of the three river types generates its own, distinctive community of floodplain plants and fisheries. Muddy rivers and even some clearwater rivers support the largest trees, but the nutrient-poor floodplains of the blackwater rivers are probably no less species-rich. Diversity is high in all types of floodplain.

There may be other, more subtle differences in floodplain forests from one end of the Amazon to the other, but we do not yet know enough to make even the most basic distinctions. In this book we will use the terms *upper, middle,* and *lower* Amazon to loosely distinguish three great sections of the main river. The *upper Amazon* is the stretch upriver of the town of Tefé, which is a thousand kilometers west of the mouth of the Rio Negro. The *middle Amazon* runs from Tefé to Santarém, at the mouth of the Rio Tapajós. The *lower Amazon* is everything downstream of Santarém. The lower Amazon thus includes the great estuarine area, which is everything downstream of the mouth of the Rio Xingu. (The lower Amazon is not to be confused with *lowland Amazonia* or *Amazon lowlands,* which describe areas below two hundred meters elevation in the main part of the basin.

In most cases the boundaries of truly distinctive ecosystems probably do coincide with geographic river basins. Unfortunately, no thorough ecological overview for even one large river basin has yet been undertaken. It is unlikely that such a survey could be carried out in less than a decade, even with adequate investment. But what *is* known leads us to one firm conclusion: the floodplain forests are the most threatened of all habitats in Amazonia. It is time, therefore, to focus on the floodplain.

During the last few years the absolute rate of upland deforestation in the Amazon Basin has decreased in many, but not in all, of its ecosystems. The cumulative damage to the uplands, or terra firme—that is, all forested areas except the floodplain—is still growing. But it now seems to be growing at a slower pace. Satellite imagery indicates that somewhere between 7 and 9 percent of the total area of Amazon rainforest has been cleared or seriously degraded. Most of the clearing occurred in the 1980s, when an average of about 20,000 square kilometers was lost every year. In the early 1990s the annual rate fell to just over 11,000 square kilometers. At the current pace the entire Amazon rainforest would be gone in just two or three centuries. The Amazon rainforest would thus follow in the path of Brazil's Atlantic rainforest, which was all but obliterated in three centuries. Today only 5 percent of the original expanse of Atlantic rainforest still has trees.

The Atlantic rainforest had evolved a high degree of species uniqueness because it was cut off from the Amazon and other rainforest areas by a drier zone. Little genetic exchange occurred between the sprawling forests of Amazonia and the sinuous, coastal forest of Brazil's Atlantic seaside. The floodplains of the Amazon are not, of course, geographically isolated from the uplands, but they are nevertheless ecologically separated to some extent because of seasonal flooding. The isolating effect of flooding has led to a high degree of endemism—comparable to that of the former Atlantic rainforest. Whereas the Atlantic rainforest follows a humid strip along the coast, the flooded forest accompanies the Amazon and its tributaries in a network of strips from the Atlantic to the foothills of the Andes.

The history of the destruction of Brazil's Atlantic rainforest is a poignant lesson in the dangers of ignoring the need for conservation and rational management of natural resources. But the looming destruction in the case we highlight in this book is not to be measured in centuries to come. It could happen in just decades. Already much of the floodplain forest is gone or vastly altered, particularly in the middle and lower Amazon. Unless action is taken within the next few years, it may be too late. The task would then be restoration, not preservation.

The flooded forests of the Amazon—along with their unique species and their vital contributions to Amazonian fisheries—are thus, in our view, the most threatened of all distinct rainforest types within the Amazon Basin.

FLOODED FORESTS AND FLOATING MEADOWS

The Amazon Basin has been the scene of great environmental changes—for a very long time. The most important geological event to affect the Amazon River floodplain in relatively recent geological history was its very creation. Much of what we now call Amazonia used to drain into the Pacific.

The Andes Mountains are a youthful fifteen million years old. Before they rose, Amazon waters flowed west and emptied into the Pacific Ocean. A northward flow might also have been an ancient pattern, with Amazon waters feeding into what is today the Orinoco River. Until fifteen million years ago, however, there was surely no eastward drainage. At that time, the Brazilian Highlands (which still mark the south rim of the basin) and the Guiana Highlands (to the north) joined within a thousand kilometers of the Atlantic. But the rising Andes eventually robbed Amazonia of a Pacific outlet. For a time, Amazonia may have been the largest lake and swamp system Earth has ever known.

A hint of that vast lake is evident every year during the wettest season, when river waters penetrate deeply into forests and replenish seasonally isolated lakes. The Amazon Basin, after all, is not called "basin" for nothing. At Manaus, which is two thousand kilometers upstream from the Atlantic, the river level is only fifty to sixty meters above sea level. Indeed, during interglacial times not long ago, when sea level was tens of meters higher, Amazonia was even more awash than it is today.

Wolfgang Junk of the Max Planck Institute calculates that the total amount of floodplain in the Amazon Basin today is at least 250,000 square kilometers. This is an area much larger than Florida—or twice that of England.

From deep gorges and rapids carved into the Andean foothills to the pounding surf of the Atlantic Ocean, the Amazon River flows for 5,900 kilometers. It courses mostly through low-lying terrain under two hundred meters in elevation. In numerous places the uplands do abut the main channel, but river cliffs seldom extend for more than a few kilometers. Most of the Amazon River is flanked, rather, by floodplains on either side. Numerous river islands are part of the floodplain, too, because of their low elevation and seasonal inundation. These river islands considerably increase the size and complexity of shoreline and inundated habitats.

During the height of the annual floods, water levels may rise as much as thirteen meters. A water depth of two to eight meters is common over much of the floodplain during this time. One can literally canoe the lower canopy at peak flood stage.

The combination of high waters and low topographic relief means that huge areas of Amazonia are seasonally inundated. The width of the floodplain on either side of the main Amazon River channel is usually from five to twenty kilometers, but in some places the floodplain may reach inland for thirty kilometers. The floodplain is thus almost always considerably wider than the main river channel—which itself is two to five kilometers wide over much of its length.

In addition to enormous size, another important feature of Amazonian floodplains is that the time of inundation lasts so long. Owing to the grand sweep of the Amazon Basin, rainfall is unevenly distributed seasonally. Peak rainfall in the northern and southern drainages is separated by several months. For example, the northern tributaries begin to rise in March or April, with recession coming in September or October. Southern tributaries commence flooding in November, with recession coming in April. This offset pattern brings flood conditions to the Amazon River as a whole for much longer than would be the case if temporal distribution were the same throughout the Amazon Basin.

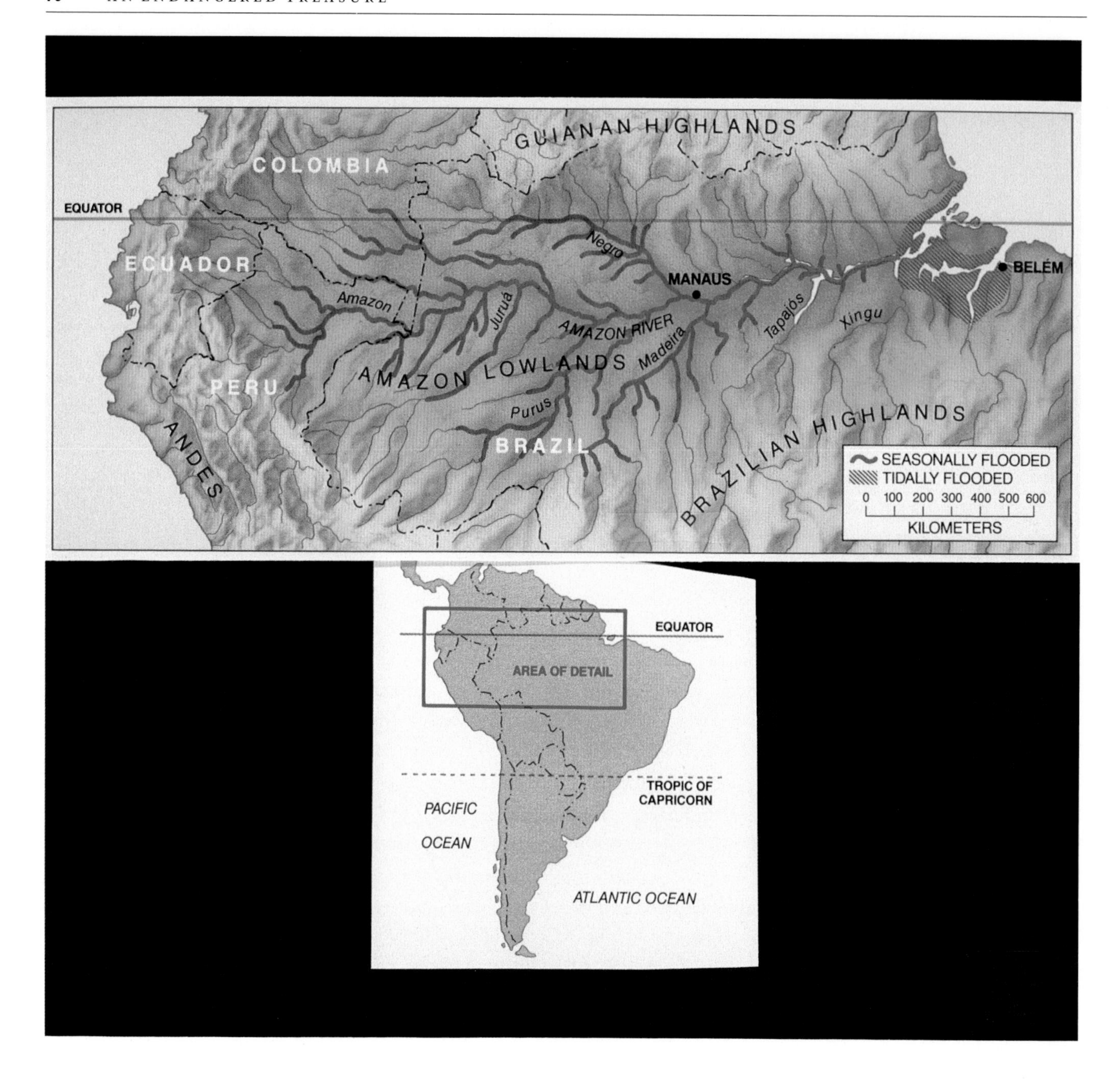

The extended flooding season along the main river and the lower courses of its tributaries is the principal factor that makes the floodplain forests as much aquatic as terrestrial habitats. And it is the flooded forests that have nurtured so much diversity and productivity of Amazon fish. *Flooded forests* in various parts of the Amazon floodplain may be inundated anywhere from three to eleven months each year, depending on local floodplain topography and the intensity of the annual floods. In general, however, the seasonal floods cover vast expanses of the floodplains for an average of four to seven months each year.

Where uplands meet the river. *Although most of the Amazon River is flanked by floodplain, sandstone outcrops do occasionally abut the main channel.*

Flooded forests are not to be confused with swamp forests, which are semipermanently or even permanently awash. Tree diversity is greatly reduced in Amazonian swamp forests because few species can tolerate being permanently waterlogged. In contrast, the seasonally flooded forests—particularly those supported by sediment-rich rivers—teem with diverse plant and animal life. Even so, it is astonishing to realize that the great trees forming a canopy over the waters may have had to spend their first ten or twenty years mostly submerged.

Flooded forests come in three types, all of which contrast with the *terra firme forests* of the upland. *Várzea forests* are fed by muddy rivers; *igapó forests* flank blackwater and clearwater tributaries; *tidal forests* mark the estuary. The tallest trees tend to grow in the nutrient-rich várzea forests. Várzea forests flank—or did before deforestation became a problem—the Amazon River itself throughout its entire course, except for the estuary.

The estuary reaches three hundred kilometers upstream, to the mouth of the Rio Xingu. One large island, Marajó, dominates the estuary. Marajó is surrounded by ar-

Canoeing the canopy. *During the several months of peak floods, one can canoe the canopy among giant bromeliads.*

chipelagoes that greatly increase the amount of shoreline and tidal habitat. The Amazon River channel discharges to the north of Marajó, but a still substantial flow also moves to the south through the Breves connection. Tidal forests differ from flooded forests upstream in that they are inundated twice daily—not just seasonally. But because of the Amazon's enormous discharge and the fact that freshwater floats atop heavier saltwater, these tidal forests are inundated by freshwater, not saltwater. Tidal forests, therefore, have largely the same stature and species as do várzea forests.

In addition to the flooded forests, the Amazon floodplain is seasonally home to *floating meadows*. These are verdant carpets of grasses, sedges, and broad-leaved floating plants, whose expansions and contractions are tuned to the rhythms of the rivers. Floating meadows appear in nutrient-rich waters fed by muddy rivers. They tend to form along the edges of riverbanks and in the shallower or wind-protected parts of lakes. While some plants in the floating meadows remain rooted in the sediments, growing vigorously upward to keep their place in the sun as waters rise, many more let their roots dangle freely. Some of the grasses, for example, can grow twenty to thirty centimeters in a day. After reaching four to six meters, however, the grasses break free of the substrate and begin to float as floodwaters continue to rise.

Plants of the floating meadows prefer open sunlight, so they grow mostly in places too deep or otherwise unsuitable for flooded forests. Floating meadows do, however, readily colonize deforested areas. Although their plant diversity may be low, floating meadows are extremely rich in invertebrates, which favor the submerged roots. Floating meadows are thus superb habitats for young fish. A single floating meadow may reach several hundred meters in width, bridging the region between flooded forest and open water for perhaps a kilometer at a stretch. The tops of the floating meadows, where especially thick, may be colonized by shrubs and vines that root in the floating organic layer.

The showcase of the floating meadows is the giant water lily. Its rimmed leaves can reach two meters in diameter; its white flowers may be as big as an adult human head. The giant water lily remains rooted, however. The root base survives throughout the dry season if the ground does not become too desiccated, while the stranded leaves and stems decay. The annual decay of plants that seasonally populate the floating meadows are, in fact, major nutrient contributors to Amazon waters.

Most of the riverbanks of the Amazon are natural *levees*, that is, ridges of sand or clay. The levees develop from the larger alluvial particles, as these are the heaviest materials in suspension and thus are the first to be deposited when rivers rise and invade the floodplains. Levees are not, however, limited to the banks of existing river channels. Many levees now arc across the floodplains—telltale signs of the borders of ancient channels that changed course. The presence of ancient levees considerably increases the complexity of the floodplain environment, or at least it does in areas where the forest still remains. Most levees along the Amazon River, alas, have lost their forests. As the highest features in the floodplain, they are favorite sites for peasant dwellings and agricultural crops.

One other habitat is characteristic of the Amazon floodplain: *floodplain lakes*. Thousands of open water bodies stud the floodplains of the Amazon lowlands. These

The showcase of the floating meadows. *The giant water lily is one of the few plants of the floating meadows that remains rooted. Its rimmed leaves can attain a diameter of two meters.*

lakes are not ephemeral; they are present year-round. But they are replenished with sediment-rich waters only during the floods, separated from the main rivers at all other times by natural levees. As with the main river channels, lakes are not attractive sites for most fish during the floods, when far more food is to be found elsewhere. Indeed, during times of low water, many of the commercially important fruit-eating fishes that inhabit the lakes or river channels are simply fasting—and avoiding predators. As soon as the floods come, they seek out the rich feeding grounds and protective cover of flooded forests and floating meadows.

The flooded forests and floating meadows are thus vital to a healthy fisheries throughout Amazonia. And it is our view that a healthy fisheries is the key to a sound and sustainable regional economy. Diversification schemes that are incompatible with fisheries, that would destroy flooded forests or floating meadows, are very unlikely, on balance, to benefit the region. For economic as well as conservation reasons, it is thus time to confront the haphazard and massive changes being brought to the Amazon floodplain.

THE HUMAN PROBLEM AND OPPORTUNITY

There are no cities on the Amazon floodplain proper. Manaus, for example, is on flood-free ground ten kilometers up the Rio Negro. But the economies of Manaus, Santarém, and Belém are vitally connected to the ecology of the floodplain.

The floodplain itself is less populated today than it once was, as more and more residents are drawn to the urban centers. Rural poverty is nevertheless still an issue. And the viability of economies in growing cities is of ever greater concern. Fundamentally, the floodplain is not an easy and prosperous place for we moderns to live. This is especially true for settlers who have come to the region out of desperation, owing to drought or famine or landlessness in the more populous regions outside

An immense floodplain lake. *The floodplain contains hundreds of very large lakes, many of which are ten or more kilometers across. The lakes are replenished by water from the main river during the flood season.*

Amazonia. Much of what the indigenous inhabitants once knew and practiced in the floodplain has been lost.

The rural peasant societies that settled along the rivers of Amazonia were a mixture of Native American, European, and African peoples. Most settlements were on the high levees bordering the main river channels, thus giving residents access to navigable waters year-round, while keeping the time of inundation to a minimum. Houses were built on stilts to put them just above the average annual flood. In some areas, indigenous peoples built earthen mounds to keep their houses safe from floods. Houses were built from locally procured floodplain timbers and palm thatch. Many floodplain houses are now, however, built with sawmill boards for floors and walls, while corrugated iron or asbestos is used for the roof. Deforestation along much of the middle Amazon is so advanced that timbers can no longer be found for house construction, and thus must be purchased in urban centers or procured in upland forests.

With the exception of propane gas stoves and radios, the inside amenities of rural floodplain houses are about as limited as they were a hundred years ago. The average house is without electricity, though floodplain villages sometimes have generators that are used during the early evening hours. Worse, most homes have no treated water or inside plumbing. Yet despite the lack of amenities, some villages on the Amazon floodplain do sport satellite dishes feeding into battery- or generator-operated televisions.

The social programs attending floodplain inhabitants are mostly a mixture of poorly funded government health and agricultural services and highly competitive religious groups that ply the rivers for converts and donors. These programs have done little to alleviate poverty or other major problems facing rural floodplain inhabitants basinwide. The average income of Amazon floodplain households is probably less than the equivalent of a thousand U.S. dollars per year.

The floodplain is not a particularly healthy place to live—especially absent plumbing and treated water. Intestinal infections are treated with herbal remedies or pharmaceutical drugs, but little progress has been made in actually preventing these diseases. Almost every aspect of human settlement on the floodplain is challenged by

Flooded out.
Every few years extreme floods drive many permanent residents from their homes.

the annual rise of the river. Food procurement becomes much more difficult during the floods, since transportation is restricted to water craft, crops are inundated, sanitary conditions around houses worsen, and heavy rains interfere with outside work.

In the past several decades floodplain peasants in many areas have turned to local and state governments for financial and other assistance during the flood season. This is especially true in times of extreme flooding, which occurs every four or five years. In these times even homes on stilts are flooded out. Given the loss of traditional expertise and the challenges presented by nature, it is perhaps not surprising that most floodplain peasants have not been able to develop enduring commercial enterprises.

The Amazon economy has been wrenched by boom and bust cycles—wild rubber and cacao in the last century and jute in the middle of this century. The current subsistence economy cannot raise significantly the standard of living of most floodplain rural societies. With the increasing urban demand for fish and other foods and products, the resources essential for subsistence are now being heavily exploited—in some cases too heavily exploited—for sale as well as consumption. The combination of increased population and livestock ranching is driving even more deforestation. Current extractive or land use activities are not likely to provide a satisfactory economic base for greatly increasing incomes over the long term.

Symbol of the flooded forest. *Fruit-eating fish (here, a pirarara catfish eating the fruit of a jauari palm) are the most unusual and commercially important denizens of the flooded forest.*

Some if not many of these activities, however, have promising components that could be nurtured and transformed into more suitable and sustainable approaches to floodplain development—while protecting the biodiversity that has captured so much global interest. Fisheries are key, of course. They are the real and sustainable fortune brought by the floods. But there are other workable (and compatible) pursuits as well.

A blend of scientific investigation and traditional knowledge systems holds the key to better use and management of natural resources. When the fruits of field-based scientific research are combined with an appreciation of socioeconomic and fiscal constraints and opportunities, more enlightened development of the Amazon floodplain should be possible. Research investment is especially needed on ways to take advantage of the annual floods rather than be drowned economically by them.

There are indeed promising ways that economic activities can be honed to make the most of the riches offered by the floods, while conserving biodiversity. It is our purpose here to explore them.

CHAPTER TWO

EARLY FORTUNE SEEKERS & THE LOSS OF NATIVE PEOPLES

Today's cultural and ecological predicament of the Amazon River and its floodplain can be understood only in the context of regional history. And that history must include the indigenous peoples who inhabited the Amazon for thousands of years before European contact.

The presence of big urban centers—notably Belém and Manaus—means that Amazonia holds more people today than it did when the first Europeans sailed up the great river. Nevertheless, there may well have been a larger human population along the Amazon River at the time of the European conquest in the sixteenth century than is represented by the total of rural settlements on the floodplain today. The Amazon floodplain has thus been extensively, if not intensively, occupied for a very long time. The indigenous human population appears to have done little harm to floodplain forests, fishes, and game animals in comparison to the devastation wrought by more recent economies.

An appreciation of the long span of human occupation of the Amazon is important on two accounts. First, it provides some indication of the rich, or at least formerly generous, resource base on which extractive activities were based. Second, a perusal of the archaeological record indicates that dense farming communities assembled long before the arrival of Europeans, thus posing the interesting question as to how they managed the natural resources. Third, the early activities of humans in Amazonia dramatically altered some plant communities through fire and enrichment planting.

FIRST HUMANS IN THE AMAZON

The Amazon Basin was inhabited long before the advent of open-field agriculture. Waves of hunters and gatherers probably entered the Amazon river system many

Pictographs.
These painted pictures on the north side of the Amazon River across from Santarém are lonely reminders of the first peoples and their ancestors.

times and from several directions. From the north, a likely early route for migrants to the basin included the Casiquiare channel, by which travelers coming up the Orinoco could cross into the watershed of the Rio Negro. From the northeast, savannas of the Rupununi in present-day Guyana and Brazil's state of Amapá could have served as gateways. Perhaps a subsequent flux of adventurers canoed upriver from the giant island of Marajó at the mouth of the Amazon.

How long people have been in the Americas, let alone the Amazon, is still disputed. Recent evidence suggests that hunters and gatherers penetrated diverse environments in the Americas at least thirty thousand years ago. Charcoal associated with hearths and possible stone tools in a cave in New Mexico have been dated to that time. A campsite at Monte Verde in southern Chile was apparently occupied thirty-three thousand years ago. Stone tools at the Pedra Furada rock shelter in the state of Piauí in northeastern Brazil are thought to be at least thirty-two thousand, and possibly thirty-nine thousand, years old. If people were in the southern cone of South America close to thirty thousand years ago, some groups were surely foraging in Amazonia even earlier.

Although some contest these early dates, other lines of investigation are converging to suggest a long history of human occupation in the Americas. Analysis of DNA of aboriginal groups in various parts of the Americas points to a possible first wave of migrants reaching the New World some forty-two thousand years ago. Linguistic evidence points to a first entry into the New World some fifty thousand or even sixty thousand years ago if only one migrant group is the founding stock for American Indian languages. Thirty thousand or perhaps forty thousand years ago is a more likely time of first entry if migrants arrived in waves.

Hunters and gatherers probably reached the Amazon very early after the entrance of humans in the New World. Cave and rock wall paintings at highly weathered Serra do Pilão, near Monte Alegre in the middle Amazon, are reputedly eleven thousand years old. Serra do Pilão is a superb site for hunters and gatherers, as it overlooks Lago Grande de Monte Alegre, a fish-rich lake so large that a traveler could easily imagine being on the open sea. Varied terrain with some patches of semi-open grasslands topping plateaus near Serra do Pilão would have provided suitable upland hunting grounds. On the floodplain flanking the rugged Serra, capybara, numerous turtles, and ubiquitous iguana surely provided the fats, protein, and calories for the early inhabitants of the sandstone caves and rock shelters.

Hunters-gatherers or incipient farmers have left their mark in other parts of the Amazon as well. Two stone arrowheads from the middle Tapajós valley are presumed to have been fashioned eight or ten millennia ago. Preceramic refuse containing percussion-flakes and remains of plants and small animals has been found in a rock shelter in the Carajás range of southeastern Amazonia; these have been dated to six or eight millennia ago. The Carajás region and the middle and upper Rio Tapajós have fewer food resources than does the biologically rich zone between the Amazon floodplain and its uplands. It is thus not unreasonable to predict that artifacts or paintings as old as twenty thousand years may eventually be found along the Amazon.

SIGNS OF INTENSIVE FLOODPLAIN AGRICULTURE

How populated was the Amazon before the devastation wrought by European diseases and conquest? And how intensively were the floodplains farmed? The records of the first European explorers provide clues.

In 1500 Europeans first witnessed the mighty Amazon, as the Spaniard Vicente Yáñez Pinzon and crew sailed 150 kilometers up the river before turning back. Another Spaniard, Francisco de Orellana, reported dense human populations on the Amazon floodplain during his expedition in 1542. He entered the Amazon River by way of the Andes, canoeing more than 4,000 kilometers downstream before reaching the Atlantic. Orellana spoke in glowing terms of the region's agricultural potential. The expedition scribe, Friar Gaspar de Carvajal, wrote of one chiefdom extending more than 200 kilometers along the river. He judged that this entire stretch of the Amazon was inhabited, with each village no more than "a crossbow shot" from the next. Further downstream, Carvajal goes on to describe the floodplain as "very plentifully supplied with all kinds of food and fruit."

Carvajal's report is important because during the following century the region was left more or less alone as early explorers reported difficult navigation, encounters with hostile natives, and little in the way of gold or other riches. But can Carvajal's words be trusted? Some modern Amazonian scholars believe not. The writings of Carvajal and other early explorers have been dismissed as gross exaggerations. Recent archaeological excavations, however, support the notion that the Amazon floodplain once supported a large population that was culturally and economically rather advanced. Vestiges of these ancient settlements, including numerous patches of so-called black Indian soils, have been found. These soils mark the now-decomposed charcoal of ancient cooking hearths and kitchen middens, and they occur widely on the uplands and floodplains along the Amazon.

High levels of phosphorus are typical of black Indian soils. Soil richness is attributed to the buildup and decomposition of fish and game bones and other discarded organic wastes. Black earth sites on the higher and more stable parts of the Amazon floodplain, such as Careiro Island near Manaus and Urucurituba near Santarém, have attracted researchers, who are now building a new understanding of the precontact

What the first Europeans saw. *The first European explorers who recorded their impressions of the Amazon wrote in wonderment of the tall trees that dominated the landscape.*

peoples. It seems that the indigenous population of Amazonia was concentrated in large settlements closely spaced along the Amazon River. In fact, the late prehistoric site under Santarém (at the confluence of the Rio Tapajós and the Amazon River) has been judged larger than the present limits of the city. The upland areas of Amazonia, in contrast, appear to have been much less densely settled at the time of conquest, presumably because of the poorer resource base.

One estimate of population density at the time of contact is twenty-eight inhabitants per square kilometer on the floodplain, and only about one inhabitant per square kilometer in the uplands. According to this view, the Amazon floodplain might have supported more than a million inhabitants. The prehistoric population density on Marajó Island at the mouth of the Amazon is estimated to have been between five and ten people per square kilometer—implying an islandwide population of perhaps 250,000. Both these estimates are, however, based only on known archaeological sites. Archaeologist Anna Roosevelt of the Chicago Field Museum cautions against this methodology. She suggests that the actual population density might have been five times as high. Overall, it is simply not possible to ascertain population density in the past based on just the surviving archaeological sites. Most vestiges of ancient villages along the floodplain have surely been destroyed by the river, which is forever reworking its channels.

Archaeological research has also shed light on the daily lives of the floodplain's indigenous peoples. Far from being culturally primitive, these prehistoric Amazonians had developed pottery making two thousand years before such appeared in the Andes or Mesoamerica. Inter-regional trade in food, clothing, and art flourished among the various chiefdoms, and large civil works projects were undertaken for water control, defense, and habitation. The population depended on hunting and fishing for animal protein. However, in the thousand years preceding the arrival of Europeans, mounting population pressures induced a marked intensification of agriculture. Subsistence seems to have been based on cultivation of root and seed crops, including maize. Fruits and nuts were also gathered from the forest. Cotton appears to have been cultivated for clothing.

Little is known about the prehistoric cropping patterns that supported large human populations, but initially sweet potato was probably the staple. Sweet potato is ready for harvest in just a few months, making it an ideal crop in an area that is under water for much of the year. Sweet potato may have been cultivated on the Amazon floodplain as long as ten thousand years ago. Another early root crop on the floodplain was probably cocoyam, since this plant is well adapted to wet places. The New World yam might have been cultivated on the higher parts of the floodplain. Interestingly, neither cocoyam nor the New World yam is important on the floodplain today. Sweet potato is still cultivated to some extent, but manioc is by far the dominant root crop today.

Manioc, thought to have been endemic to northeastern Brazil, was probably introduced later to the Amazon floodplain. It could have been planted easily on the higher parts, which may flood only a few times each decade. Rapidly maturing manioc varieties would have been favored. Centuries of cultivation would thus have led to

Harvesting maize as floodwaters rise. *Maize was one of the principal crops that Europeans found in Indian villages. After the Indians were decimated, maize was largely replaced by the root of manioc. Once again floodplain farmers are beginning to experiment with maize, and it is a shame that the diversity of floodplain maize types that Indians once had no longer exists.*

varieties increasingly adapted to the large area of low-lying floodplain that could be planted each year when waters receded.

Given the plentiful supplies of fish, along with turtle, manatee, capybara, and several species of ducks, heavy reliance on starchy root crops would not have provoked protein malnutrition. Vegetable sources of protein would also have supplemented the diet. We know that extensive floating pastures of wild rice were harvested by shaking the ripe panicles into canoes or baskets. On the island of Marajó at the mouth of the Amazon, some of the wetland grasses may have been semidomesticated.

Maize, or corn as it is commonly called in the United States, arrived in the Amazon very early. A highly productive and nutritious cereal, maize soon was introduced from its area of origin in Mexico and Guatemala, apparently reaching the Napo River in western Amazonia by six thousand years ago and the middle Caquetá perhaps a thousand years later. Maize was then brought down the Amazon by various indigenous cultures. It was both eaten and, after fermentation, drunk. Orellana's expedition encountered a village along the Amazon between the Rio Negro and Rio Madeira that was distinguished by two massive tree trunks carved in the form of two jaguars. These structures supported a tower with a central font containing maize beer.

Many indigenous groups probably farmed both the alluvial soils and adjacent uplands, as numerous communities on and around the floodplain still do today. In this manner, people draw upon a great diversity of crops for subsistence and trade. Pollen analysis of cores from some of the more permanent lakes on the Amazon floodplain may eventually reveal the exact sequence of crops cultivated on the floodplain.

Apart from the muscovy duck and possibly some stingless bees, no truly domesticated animals appear to have been kept by indigenous peoples in Amazonia. Abundant sources of animal protein along the rivers, especially silt-laden rivers such as the Amazon, may have dissuaded groups from taking the trouble to tame, house, and feed animals. Many mammals and birds were kept as pets, but these were usually captured in the wild while still young. Some "pets" were eventually eaten, but others, particularly parrots, macaws, and harpy eagles, were kept to supply feathers for ornamental wear. At the time of conquest, Indian groups kept thousands of South American river turtles in aquatic corrals along the Amazon, but these were captured after laying eggs on beaches at low water, rather than raised in captivity.

Large populations of farmers and fishermen along the Amazon floodplain could have been supported with less forest clearing than occurs today. Indigenous groups undoubtedly cleared some of the floodplain forest to build homes and to plant crops, but they did not have cattle or water buffalo—a major factor in current deforestation of the Amazon floodplain. Another reason why indigenous people probably cleared only limited areas of the floodplain is that their garden plots were undoubtedly highly productive. Furthermore, they needed and valued forests for a variety of products, from timber for construction and canoes, to fruits, nuts, and medicinal plants.

Another reason why floodplain forests were unharmed by relatively dense settlement in prehistoric times was that buffer zones were apparently established between some warring groups. Along the central Rio Ucayali in the Peruvian Amazon, for example, a buffer zone separated the Conibo and Cocama tribes during the sixteenth and early seventeenth centuries. On the upper Amazon, buffer zones of varying lengths helped minimize contentious contacts between such well-known warriors as Aparia the Great, the Machiparo, the Omagua, and the Paguana.

Such de facto buffer zones would have left floodplain ecosystems essentially intact, at least as long as hostilities persisted. Even though some buffer zones surely collapsed as groups made peace or were replaced by more powerful or organized societies, other off-limit zones would eventually materialize. The length and duration of such buffer zones is uncertain because the floodplain of the middle Amazon is more than twenty kilometers wide in some areas; early explorers may have missed villages along the interface between floodplain and uplands, or along the margins of interior lakes of floodplain islands.

CONQUEST, DISEASE, AND SLAVERY

As everywhere in the Western Hemisphere, the indigenous population of Amazonia plummeted after contact with Europeans. Missionaries and explorers brought with them a host of pathogens for which the indigenous populations had no resistance. Smallpox, influenza, measles and perhaps tuberculosis diffused quickly among the dense populations. People attempting to flee the new pestilences only made matters worse by spreading sickness further afield. In some parts of Latin America, as many as 90 percent of the Indians died within a century of contact with the Spanish and Portuguese.

The portion of the indigenous population that perished along the Amazon during the first hundred years after contact with Iberians is unknown, but it must have been substantial. In 1689, for example, Friar Samuel Fritz saw no people or settlements during six full days of travel downstream of where the Rio Urubu joins the Rio Amazonas. In contrast, Francisco de Orellana and his crew had been besieged by hostile natives when they navigated a similar stretch of the Amazon, downriver of the Rio Negro, a century and a half earlier.

The Portuguese established only a few small towns along the Amazon. Massive destruction of indigenous societies, abetted by slave raids, continued throughout the colonial period. In hope of eluding slave traders, Jesuit and Franciscan missionaries encouraged remaining indigenous groups to move away from the Amazon floodplain to more remote locations.

The Treaty of Tordesillas, negotiated in 1494 by Pope Alexander VI, attempted to carve up the world between Spain and Portugal. Under this agreement, a demarcation line was drawn 370 leagues west of the Cape Verde islands in the Atlantic, just off the westernmost point of Africa. All lands to the west of this line were to belong to Spain and all territories to the east to Portugal. The apparent intention of the pope was to give Africa and Asia to the Portuguese and the newly discovered Americas to the Spaniards. Although there was some dispute over the exact location of the Tordesillas line, it was generally agreed that it passed slightly to the east of the main mouth of the Amazon River. Officially, the Amazon region began as a Spanish colony.

Although Orellana's epic journey down the Amazon River generated widespread interest, only a few abortive attempts were made by the Spanish to colonize the region. By the early 1600s, the main contenders for control of the Amazon were the Portuguese, Dutch, and English, all of whom had established outposts at various points along the river. In 1616 the Portuguese, in violation of the Treaty of Tordesillas, built a fort south of the mouth of the Amazon River at what is now Belém, from which they successfully expelled the other foreign interlopers. Another fort was built in 1669 by the Portuguese at the confluence of the Rio Solimões (Amazon) and Rio Negro, a site that has become the sprawling city of Manaus. Manaus, named after an extinct Indian group, became the first major European population center in the interior of the Amazon.

With de facto control of Amazonia now secured, the Portuguese set about colonizing the region and taking advantage of the natural wealth so vividly described by early explorers. One of the first necessities of the new European settlers was to secure an adequate labor supply. Most of the colonists were reluctant to engage in manual labor, even those who had been farmers in Europe. Too poor to purchase African slaves, they quickly resorted to exploiting the only labor force available: the indigenous population. Indians became the "red gold" of the Amazon. As described by the Royal Geographical Society's president, John Hemming, "during the first decades after the Portuguese arrived on the Amazon there was an almost unrestrained 'open season' on its inhabitants."

Indian labor was secured in two ways. The main method was through slaving expeditions that either purchased Indians already held captive by other tribes or seized them from their own villages. The Indians, decimated by disease and in a state of cultural collapse, were no match for even the few hundred Portuguese equipped with firearms. The second way was to have missionaries, mainly Jesuits, persuade entire tribes to settle in compounds called *aldeias*. Indians residing in mission villages located near European settlements were obliged to work for the colonists at pitifully low wages. The Crown also reserved the right to requisition mission Indians to construct public works, paddle canoes, and to participate in slaving expeditions.

The enslavement, disease, and cultural disruption of the riverine Indian population caused a complete collapse of the intensive, preconquest form of floodplain agriculture. For a number of reasons, the Portuguese were never able to replicate the highly productive and sustainable agricultural practices of the original inhabitants of the Amazon floodplain. Traditional European cultivation patterns generally did not fit well with the physical characteristics of this strange new land. And lost forever was the complex social organization that had permitted the Indians to exploit the floodplain so effectively for thousands of years. Belém, by far the largest European population center in the north of Brazil, had only five hundred inhabitants in 1700.

EXPORTS BASED ON WILD PRODUCTS OF THE FOREST

Eschewing sedentary agriculture, most colonists concentrated their efforts on extracting exportable forest products, the so-called drugs of the backlands, from both the floodplain and upland rainforests. The most coveted of these products were the root beer vine (used for making medicinal teas), false clove (a tree whose leaves have a clovelike aroma), false cinnamon, vanilla, and cacao. Sap from the copaíba tree and oil from the andiroba nut were also valued for their use in pharmaceutical products, soaps, and lamps.

Collecting missions commonly lasted six to eight months. Expenses for these missions were minimal, especially compared to those of plantation agriculture. Typically, expeditions consisted of one or more large canoes, each capable of transporting as much as eight metric tons of cargo, with a crew of one to two dozen Indian paddlers. The Indians were commanded by a white or mixed-blood boat master. The success of the venture depended on the Indians' intimate knowledge of the Amazon forest and

Medicinal plants. *When the early Europeans failed to find large quantities of gold in the Amazon, they turned to forest products, and especially plants of medicinal value. Expeditions explored the interior for these products. Even today, medicinal plants are still sold in Amazonian markets.*

thus their ability to locate the desired products. The Crown controlled the forest products business by requiring each expedition to obtain a license before heading upstream. By the early 1730s, about 250 such licenses were issued annually. Upon the return of each mission, colonial authorities weighed and taxed the products collected.

More than sixty Jesuit-managed mission villages had been established along the Amazon River by the mid 1700s. As many as fifty thousand Indians were concentrated in these villages in peak years. In contrast to those kept by colonists for slave labor, mission Indians were legally "free." Religious orders also strove to keep Indian families together and to group them with members of their own tribe. While Jesuits were known to be good managers, their humane treatment of Indian workers was probably the principal reason why missionaries did better at agriculture and other economic activities than did the colonists.

A typical mission village supported subsistence gardens based primarily on manioc. Surrounding the villages were often small plantations producing cotton, tobacco, and introduced crops such as sugarcane and rice. The Jesuits also owned several large ranches on the island of Marajó at the mouth of the Amazon. These ranches reportedly contained some thirty thousand head of cattle. In addition to these enterprises, the Jesuits mounted collecting missions for the various "drugs of the backlands." Sales of surplus produce from these activities were handled by lay agents in Belém and São Luis. The proceeds were returned to the village to help with general upkeep. The relative prosperity of the religious orders over time, however, inspired serious resentment on the part of the colonists.

No statistics on regional exports prior to 1730 are available. The Amazon, however, was almost certainly the poorest part of Brazil during the colonial era. During the first quarter of the seventeenth century, for example, only one or two ships per year called at the port of Belém, versus twenty to forty ships at major port cities in northeastern and southern Brazil. A modicum of wealth was achieved in the early 1700s when a growing taste for chocolate drinks in Spain and other European countries contributed to rising demand and prices for cacao. Though it may seem surprising today, cacao was the Amazon's major export product during the colonial era and remained so well into the 1800s. During the 1730s and 1740s, for example, cacao in most years accounted for 90 percent or more of the region's total exports.

To meet growing external demand during the last decades of the seventeenth century, the Crown tried to promote development of cacao plantations in the Amazon. Demonstration orchards were established near Belém, and the ban that had prevented government officials from exporting cacao on their personal accounts was lifted. Fiscal incentives, including full or partial exemptions from customs duties, were also put into effect. But these early efforts met with little success.

Plantation-grown cacao did not take off in the Amazon during the colonial era for a variety of economic reasons. One key factor was that wild cacao grew profusely on the floodplain of the main river, especially between Belém and the upstream towns of Óbidos and Santarém. It could also be found easily along the banks of some of the Amazon River's major tributaries, such as the Negro, Trombetas, and Madeira. In contrast, planted cacao took five years to reach maturity and thus tied up financial resources that otherwise could have been used for financing collecting missions. Finally, there was the perennial Amazonian problem of labor shortages, which were particularly acute for plantations, as Indians did not adapt well to the regimented manual labor required by this type of agriculture. Unit labor costs must have been very high for plantation owners.

In 1750 the Treaty of Madrid was signed by Spain and Portugal. It formally recognized Portugal's claim to most of the Amazon region. Despite early successes in the cacao trade, however, northern Brazil was in dreadful condition. The chief cause of this decay was the near annihilation of the region's Indian population, the main source of labor. Travelers along the main river in the mid 1700s found the banks virtually devoid of human life. Although this plunge in population was in some cases due to mistreatment of Indian slaves by the colonists, the main culprits were epidemics of imported diseases—measles, influenza, and smallpox—to which the indigenous population had no resistance. It is not known how many Indians perished in Amazonia during the first century of Portuguese rule, although the total must have been in the hundreds of thousands. For example, forty thousand people were reported to have died in the Belém area alone during the smallpox epidemic of 1750, and the vast majority of these were undoubtedly Indian.

The precipitous demographic and economic decline of Amazonia in the mid 1700s accentuated the long-standing rivalry between the colonists and the Jesuits. Animosity stemmed mainly from jealousy over the Jesuits' relative prosperity, although the colonists were also irritated by the religious order's protection of the Indians. These attitudes gave rise to a series of new laws aimed at reducing the influence of the Jesuits, while at the same time reviving the regional economy.

In 1755 colonial authorities issued the "Law of Liberties," which released all Indians from slavery and granted them political self-government. Closely following this seemingly altruistic law were other edicts wresting control of mission villages from the Jesuits, confiscating their worldly goods, and eventually expelling them from Brazil. White "directors" were put in charge of the former mission villages (now "directorate" villages), and elaborate rules regarding the social and economic obligations of the Indians were put into place. Because of the way directorate villages were managed, the law giving freedom to the Indians would prove to be a sham.

In an additional move, the Portuguese established a new trading company—the General Company of Commerce of Greater Pará and Maranhão—aimed at stimulating agriculture in the region. This was to be done by importing African slaves to work on the colonists' plantations, by making available imported merchandise at reasonable prices, and by providing shipping facilities.

During the later part of the eighteenth century, colonists and directorate villages both engaged in the collection and export of natural forest products. A small agricultural sector also started to evolve. The colonists, some of whom were now supplied with African slaves, concentrated their efforts on large-scale production of exportable crops, such as sugarcane and rice. The directorate villages, in contrast, focused on food crops like manioc, maize, and beans to satisfy their own needs as well as to sell in Belém and in the military outposts.

Directorate villages functioned under a strict command-and-control system. Indians were told what to plant, how much to produce, and how to allocate their time. Half of the able-bodied men in the villages were made available to the colonists as wage labor.

According to some scholars, all these social and economic changes made in the post-Jesuit era substantially revived the regional economy. The data on regional exports for the period, however, do not support such a sanguine view. During the short tenure of the General Company (1755–1778), exports from Belém grew only modestly. A brief export boom, which started in 1790 and ended in the early 1800s, seemed to have little to do with the new laws and policies. It was probably more related to rising prices for cacao, stimulated by the Napoleonic Wars, which cut off supplies of this product from Spanish Venezuela.

By all accounts, the economic impact of the directorate villages was minimal. The social impact of the system on the Indians, however, was disastrous. They were paid only a pittance for their labor, their culture was demeaned, and vicious beatings were meted out for even minor infractions of the rules set by the village directors. In light of this treatment, it was not uncommon for village Indians to flee into the forest. Disease epidemics also continued to claim large numbers of victims. At the height of the directorate system in 1772, fewer than twenty thousand Indians were under Portuguese control in Pará. Of these, only four thousand were in the villages. Indian labor had simply ceased being a major economic factor in the Amazon region by the end of the eighteenth century.

By the end of the colonial period, many Indians uprooted from their tribal ancestry, along with people of mixed blood known as *caboclos*, had spread out along the rivers in highly dispersed settlements. In sharp contrast to the situation prevailing at the time of the European conquest, these people came to rely more and more on the relatively infertile soils of the terra firme, or upland forest, for their subsistence needs. The widespread adoption of manioc as the primary cultivar made this switch to the upland possible. Furthermore, the Portuguese, the Spanish, and their *caboclo* descendants lacked a knowledge of floodplain agriculture, and thus turned mostly to the uplands.

The fertile floodplain could only be cultivated during the dry season, when the river was at its lowest. This period, however, coincided with the peak season for gathering forest products, the main source of cash income for the riverine population. Owing to prevailing labor shortages, it became virtually impossible to practice both activities concurrently. Individuals also tended to travel widely in search of "drugs of the backlands," an additional factor militating against the practice of intensive floodplain agriculture.

Some export crops were grown on the floodplains during the decades immediately following independence in 1822. Cacao groves, for example, were cultivated and harvested regularly on the riverbanks near the towns of Cametá and Santarém during the mid 1800s. Judging from contemporary accounts, however, these enterprises do not seem to have been very prosperous. By and large, the regional economy depended on the export of various natural products harvested in the wild. This point is clearly illustrated by the trade statistics of Amazonas Province for 1853–1863. Tobacco was the only cultivated crop to appear in the provincial trade statistics of that period, and it is of relatively minor importance. Of much greater significance are exports of "drugs of the backlands," such as the root beer vine and copaíba oil, and products from the river, such as the pirarucu fish (in salted form) and the oil rendered from turtle eggs.

Except for the gathering of the root beer vine, which killed the plant, the environmental impact of extracting forest products seems to have been relatively benign. The same cannot be said, however, for the harvesting of aquatic life—especially for the take of turtles and manatees. Fishes and aquatic reptiles and mammals have been a focus of exploitation since the first European contact with the region. The colonial civil and religious authorities apparently tried to set limits on the harvest of turtle eggs and meat, but enforcement was a problem. During the early 1700s an estimated twenty-four million turtle eggs were harvested annually in the upper Amazon to produce oil. By the 1850s the annual harvest of eggs had doubled.

This reckless exploitation of aquatic life took its toll—on the economy as well as the resource. As early as the mid 1800s, observers were already commenting on the growing scarcity of river turtles in the Amazon region. Manatees resisted the pressure a bit longer, but they too became more difficult to find as the century came to a close. The ruin of these major types of aquatic resources was perpetrated by a relatively small number of individuals. The total population of Amazonia was only two hundred thousand in 1850. In contrast, it seems that fishes, turtles, and manatees abounded at the time of conquest, even though a much larger human population relied on these creatures as the main source of animal protein. This suggests that Indian tribes had established taboos against the overexploitation of wildlife or had other effective cultural checks that were unwisely discarded by the European conquerors.

THE COMING OF THE RUBBER TAPPERS

Rubber was the single fastest-growing export of the Amazon between 1853 and 1863. From the mid 1800s to the end of World War II, the economic well-being of all of Amazonia was inextricably linked to the strength of world markets for this product. Several species of wild rubber are native to the Amazon Basin. But one species, *Hevea brasiliensis*, was by far the most important for commercial use. This thirty- or forty-meter tree is found in areas south of the Amazon River. Rubber trees, as a group, are found in both the upland and floodplain rainforests. The larger, more commercially valuable trees, however, are located in upland areas with well-drained soils. Because of easy access, most of the Amazon floodplain, from its westernmost reach all the way downriver to as far as Santarém, was exploited as well. Rubber trees are distributed sparsely in their natural state, with only two or three individuals per hectare being the norm.

Long known by Indian tribes, latex extracted from rubber trees came to the attention of the Portuguese colonial authorities around 1750. From time to time, boots, knapsacks, and other articles were actually sent from Lisbon to Belém for waterproofing with the impervious product. A small export market also developed as word of

Turtle eggs for food and oil.
The giant Amazon river turtle became a favorite of European colonists. During the early 1700s an estimated twenty-four million turtle eggs were harvested each year. This biggest of all Amazon turtles has largely vanished from the main course of the Amazon River.

Rubber tapper.
The rubber boom at the end of the last century attracted many settlers from poverty-stricken northeastern Brazil. Rubber tappers and their families spread out along Amazonian rivers, and their descendants today make up a large part of the riverine culture in Brazil.

this new product spread. By 1800, for example, New England merchants were placing orders in Belém for rubber shoes. A major boost was given to rubber exports by Charles Goodyear's discovery of vulcanization in 1839, a process that stabilized rubber and prevented it from melting in the heat and cracking in the cold. The advent of steamboats on the Amazon River in 1853 also facilitated exports by reducing the travel time for the Belém-Manaus-Belém run from several months to only twenty-two days. Steamboats also released labor, always a scarce commodity in the Amazon.

Between 1850 and the peak year of 1912, exports of wild rubber from the Amazon increased more than twentyfold—from 1,500 to more than 31,000 metric tons. Rubber rapidly supplanted cacao as the region's leading product and, in the early 1900s, seriously challenged coffee (grown primarily in the South) as Brazil's leading export. To meet the growing world demand for latex, hundreds of thousands of men were recruited to work as rubber tappers. Most were drought victims from the arid interior of the Brazilian Northeast. The rubber tapper recruits and their families spread out along the rivers, thus repopulating the floodplains for the first time since conquest.

With world prices rising to as high as three U.S. dollars per pound by mid 1910, the resulting economic boom was bringing an undreamed of prosperity to the region.

Manaus, the center of the rubber trade, grew from a backwater village of only five thousand inhabitants in 1870 to a cosmopolitan city of fifty thousand in 1910. Urban amenities included electricity, telephones, efficient water and sewerage systems, a magnificent opera house, and palatial mansions for the new rubber barons. Owing to shortages of labor and a general aversion to agriculture, almost all consumer goods were imported, even basic food items such as rice and beans. Prices of virtually everything in Manaus were said to be four times those in New York City. Although the high cost of living was undoubtedly a burden on the poor, the rich did not seem to be concerned. In 1910 alone, Manaus high society imported 729 cases of champagne, 4,223 cases of whiskey, and 1,594 cases of perfumes.

The huge profits generated by wild rubber exports inevitably attracted the attention of potential foreign competitors. The British, in particular, were anxious to explore the possibilities of growing *Hevea* on plantations in their Asian colonies. To this end, Henry Wickham, a young Englishman residing in Santarém, shipped 74,000 *Hevea* seeds to the Royal Botanical Gardens in Kew in 1876. There is still some debate over the legality of Wickham's shipment. However, the contention that he "smuggled" rubber seeds out of Brazil is probably overstated, since rubber seeds had been exported from Amazonia previously with the full knowledge and consent of the authorities. Wickham himself apparently fueled the "seed snatch" myth in an attempt to demonstrate his bravado back home. Whatever the case, within a few decades of Wickham's shipment, the British had established plantations in their Asian colonies that could produce rubber at only a quarter of the cost incurred by Brazilian producers.

Plantation-grown rubber started entering world markets on a grand scale in late 1910, and by the end of the following year the price of rubber had dropped to a dollar per pound. One scholar estimates that there was, at the time, more *Hevea brasiliensis* growing on Asian plantations than growing wild in all of Amazonia. The Brazilian government made a futile attempt to save the local rubber interests through the so-called Rubber Defense Plan. Cash premiums were given to people who planted rubber trees and built rubber-processing plants, export taxes on rubber were drastically reduced, and import taxes on inputs to the rubber industry were abolished. These official attempts to bolster the regional economy did nothing to stanch the ever-increasing flow of Asian production, however, and prices plummeted to barely fifty cents per pound by 1914.

Between the end of the rubber boom and World War II, Amazonia suffered through three decades of stagnation and despair. Real income in the region is estimated to have declined by 80 percent between 1910 and 1920. This prompted many rubber tappers to return to their native northeast. Census data show practically no population growth in the Amazon region between 1920 and 1940, and estimates of net emigration for this period are in the neighborhood of two hundred thousand people. Those who remained in most cases eked out a meager existence through subsistence agriculture, fishing, and the extraction of forest products. Many people living in Amazonia today are, in fact, descendants of these former rubber tappers.

The British formed a cartel during the early 1920s in an attempt to restrict the world supply of rubber and thus raise commodity prices. In a defensive move, the Ford Motor Company decided to establish its own rubber plantations in the Brazilian Amazon. One million acres were purchased in 1927 by Ford on the east bank of the middle and lower Tapajós River. Seven years later the company established a smaller plantation on a low plateau just fifty kilometers south of Santarém. It was hoped that the higher and drier location of the newer operation would protect rubber seedlings from disease. Both plantations gave rise to support communities that came to resemble small North American towns, complete with hospitals, schools, movie theaters, water, and electricity.

The Ford plantations suffered serious problems throughout their existence, including shortages of workers, inappropriate planting materials, and severe infestations of the South American leaf blight, which swept through the vulnerable monocultures. All these problems, particularly the leaf blight, reduced output and raised costs. Ford was able to squeeze a mere seventy-five kilograms of latex per year out of each hectare. This was far less than the takes achieved by smallholders in Asia during the 1930s. Production costs, moreover, were an extremely high thirty-five cents per pound. By 1945 mounting financial losses prompted Ford to sell the two plantations to the Brazilian government for a quarter million (U.S.) dollars. Company records indicate that the Amazon rubber plantations had cost Henry Ford almost thirteen million dollars, most of which was lost.

Entry of the United States into World War II provided temporary relief to Amazonia from decades of stagnation and neglect. Through the "Washington Accords" of March 1942, Brazil agreed to cooperate with the Allied forces in supplying strategic raw materials, including rubber. A major effort to mobilize production began. To finance this undertaking, the Import-Export Bank offered Brazil a credit of one hundred million dollars.

The agreements underlying this so-called Battle for Rubber called for Brazil to export all rubber in excess of domestic needs to the United States at a fixed price of thirty-nine cents per pound, which was to be raised in steps to sixty cents in 1944. An added inducement for Brazil's redevelopment of its rubber industry was the offer of cash premiums to be paid on exports annually exceeding five thousand tons. To comply with these production commitments, the Brazilian government was faced with the immediate problem of mobilizing a labor force. A special agency was set up in 1942 to assemble tens of thousands of workers in Belém. Between 1942 and 1945, this agency transported more than thirty-two thousand workers and their families to Amazonia. Once again a mass migration was facilitated, as it had been during the 1870s, by a prolonged drought in the northeast.

In relation to the effort and costs involved, the production results of the Battle for Rubber were quite modest. From 1941 to 1945, the annual output of wild rubber rose from fifteen thousand to twenty-five thousand metric tons. The 1945 production was only slightly more than half that registered in the peak production year of 1912. Furthermore, because of the short-term "emergency" nature of the program, its impact on the region was transitory.

In the decades following European discovery and conquest, the vast stretch of the Amazon floodplain was quickly transformed from a thriving agricultural region of indigenous chiefdoms to a depopulated and deculturated wilderness with just a few village centers. Early exploitation of plants by the Portuguese for export markets effected minimal harm to the forests for the next several centuries, but valued animals were soon overharvested—notably turtles and manatees. Manaus became a city only during the rubber boom of the late nineteenth century, and a flux of immigrants then repopulated the floodplain to pursue the new trade of rubber tapping. The rubber boom was followed by the rubber bust; rubber tappers returned to their native northeastern Brazil or stayed on in a subsistence mode. Still, the flooded forests and floating meadows of the Amazon remained much as they always had been. The relatively benign coexistence of humans and nature would not, however, last.

CHAPTER THREE

BOOM & BUST IN MODERN TIMES

The past four decades have wrought more environmental change on Amazonian rivers and floodplains than had occurred during all of Indian and colonial history. Most of the economic activities at the root of such change have had little impact on world markets, but their regional environmental consequences have been immense. The deforestation of natural levees for jute farming and the almost complete logging-off of the Amazon floodplain are two cases in point.

The international banking system provided relatively easy loans to the Brazilian government in the 1970s and early 1980s, and this allowed the launching of several megaprojects aimed at rapid development. Of these projects, dams were the most ambitious. But the huge foreign debt incurred by Brazil, coupled with national and international concern for the environment, has thus far thwarted most of the grandiose schemes proposed.

Unlike the hydroelectric potential of Amazonia, timber and precious metals are conducive to small-scale and private exploitation. The miniboom in floodplain logging began in the 1960s, followed in the 1980s by a massive expansion in the interior of Amazonia. By then, most of the floodplain had been stripped of its most valuable tree species, and attention turned to the tidal forests. The late 1970s thrust upon Amazonia one of the greatest gold rushes ever in the Western Hemisphere. Gold fever continues to grip the region. The thousands of isolated mines, which operate largely outside of the formal economy, have brought mercury pollution to Amazonian rivers. Almost nothing has been done to stop it. Finally, cattle and water buffalo are now being placed on the floodplain in large numbers. This kind of ranching is destroying the flooded forests and floating meadows. Floodplain ranching will be discussed in chapter 8. Here we outline the story of high-impact agriculture, mining, logging, and hydroelectric projects.

Jute on its way to market. *Asian jute, a fiber crop, launched the first widespread agricultural development of the Amazon floodplain. Most of the jute factories have now closed, as their product cannot compete with synthetic materials used for packaging and rope.*

RISE AND FALL OF THE JUTE BUSINESS

The first high-impact, commercial crop grown in the floodplains was jute. Jute is an exotic plant of the linden family, endemic to Asia but grown as a fiber crop throughout the tropics. During most of the post–World War II period, the agricultural sector of the Amazonian floodplain was dominated by the production of jute.

Jute was first cultivated in Brazil just after World War I, in the southern state of São Paulo. The results of this experiment were not encouraging, as production was low and labor costs high. But in 1929 Japanese immigrants began jute cultivation on the floodplains near the city of Parintins in the state of Amazonas. Here they encountered ideal soil, water, and climatic conditions. Production techniques were perfected by the mid 1930s, and cultivation quickly spread to the floodplains of Santarém, Óbidos, Alenquer, Monte Alegre, and Juruti in the neighboring state of Pará. The total annual output of jute from Amazonas and Pará expanded from 80 metric tons in the late 1930s to almost 25,000 metric tons in 1953, the year Brazil achieved self-sufficiency in jute production. Many cacao groves and some floodplain forest were cleared to make way for this profitable crop.

The production of jute on the Amazon floodplain was, and continues to be, a smallholder activity. It is a backbreaking and unhealthy occupation, however, requiring workers to remain waist-deep in water for long hours. Typically, jute producers cultivate plots of two or three hectares. Seed is spread by hand on the floodplain's natural levees in December to January in the lower Amazon. This is the beginning of the rainy season. The harvest comes three to four months later, before the floodwaters cover the crop. As the waters penetrate jute fields, the stems are cut and soaked so that the fibrous component can be easily stripped off the bladelike stems.

Jute production is primarily a family enterprise, although wage labor may be employed to help with the harvest. Fertilizers and other purchased inputs are almost never used, and about the only out-of-pocket expense is the purchase of seeds. Since jute farmers do not normally hold legal title to the land, they frequently have a difficult time obtaining credit from the formal banking system. Such credit is most often provided—usually in the form of basic food items like sugar, rice, and cooking oil—by urban commercial interests, such as seed merchants or jute-processing companies.

Most jute is sold on the domestic market. Although jute farmers have received little direct assistance from the government, they have benefited indirectly from the national policies of import substitution and export promotion pursued by Brazil since World War II. The main use of jute, often in combination with a native Amazon fiber known as *malva,* has been the manufacture of sacking used to store agricultural products destined for export—such as coffee, sugar, soybeans, and cacao. The local value of jute thus swings with the fluctuations in international commodity markets.

The heyday of Amazonian jute came during the 1960s. Jute accounted for perhaps a quarter of the value of all agricultural production in the Amazon region during the mid 1960s. Twenty thousand families were then engaged in jute production in the state of Amazonas, and another six thousand in Pará. During the 1970s, competition from plastic sacking began to cut into the jute market. Petroleum-derived polypropylene bags are today sold in Brazil at half or even a third of the price of those made from jute, and these less expensive containers are now widely used for many commodities, particularly for groundnuts, potatoes, and onions. A second factor in the decline of jute is the shift to bulk handling of many commodities, especially sugar, grains, and soybean. Finally, producers in Bangladesh have driven down the price of jute fiber because of their relatively low operating costs. Bangladesh now provides close to half of all jute traded on world markets.

One symptom of jute's decline in Brazil is the shrinking number of factories that process the fiber. Manaus once had seven jute factories, but only two were still operating in the early 1990s. All three jute factories in Santarém have shut down, the last one in 1986. Juriti, once a major hub for jute production, no longer has any factories for pressing and baling the stringlike fiber.

The time of jute farming along the Amazon has thus come and gone. In 1991 the Brazilian government eliminated tariffs for jute imported from Asia. The remaining demand for jute used in Brazil is thus likely to be satisfied by producers in Bangladesh. The only natural fiber plant with some prospects for increased cultivation on the floodplain is sisal. Native to the drier parts of Mexico, sisal is grown by a few floodplain farmers in the vicinity of Santarém, where it is used locally for high-quality rope. Sisal producers on the Amazon floodplain are, however, unlikely ever to outproduce growers in such countries as Tanzania and Kenya.

OPERATION AMAZONIA

Four centuries after its discovery by Europeans, the Amazon was still essentially isolated from the rest of Brazil. This isolation was largely due to the great distances involved and the lack of overland transportation links with the dynamic south of the country. In 1960 the entire Amazon region had only 6,000 kilometers of roads, less than 300 kilometers of which were paved. The Amazon River and its tributaries, by necessity, continued to function as the main avenue of access to the vast hinterland.

Lacking a highway network, the regional population centers and most economic activity remained clustered along major waterways, as they had been since colonial times. The extensive tropical rainforests elsewhere were virtually devoid of human life except for a few scattered Indian tribes. As a result, disturbances of the terrestrial environment were modest and largely confined to the clearing of floodplain forests close to urban centers for timber and agricultural purposes. The physical isolation of Amazonia, and the protection this provided to the upland rainforests, came to an end, however, in 1964. A highway 1,900 kilometers long was completed, connecting the new national capital of Brasília with Belém, which is near the mouth of the Amazon River.

The move of the national capital from Rio de Janeiro to Brasília in 1960 and the subsequent development of an inter-regional highway system were both part of a major government effort to open up the interior of the country for settlement and development. The Brazilian military, which took over the national government in 1964, quickly demonstrated a keen interest in the development of Amazonia. In a series of legislative acts and decrees enacted in 1966 and 1967, cumulatively known as Operation Amazonia, the new government firmly committed itself to occupation of the northern region and exploitation of its resources, as well as the eventual integration of the North with the rest of Brazil. These plans included an ambitious road-building program to link Amazonia with the Northeast and South, agricultural colonization schemes, and fiscal incentives to attract new industrial and agricultural enterprises.

The motives behind Operation Amazonia were largely geopolitical. Several neighboring countries had already initiated programs to occupy and develop their own Amazon regions, and Brazil's military leaders were anxious to ensure national sovereignty by establishing self-sustaining settlements in frontier areas. Nationalist factions had also become alarmed by a proposal by the U.S.-based Hudson Institute to dam the Amazon, as well as by revelations that large tracts of land in the region were being acquired by foreigners. Another motive for opening up the region was the notion that Amazonia could serve as a safety valve for demographic and social pressures building up in the chronically poor Northeast. Land settlement schemes aimed at the new highways were to provide, according to a popular slogan of the era, "land without men for men without land."

The road-building component of Operation Amazonia was accomplished with dispatch, although a highway designed to parallel the northern bank of the Amazon River was abandoned for technical and financial reasons. By 1980 the regional road system had grown to 45,000 kilometers, a sevenfold increase in just twenty years. For the first time in history, all three of the major cities located on or near the Amazon River—Manaus, Belém, and Santarém—were connected to southern Brazil by overland means. The Manaus and Santarém links, however, have been very poorly maintained and today are largely impassable.

In contrast to road building, the results of the directed settlement programs were extremely modest. By the end of the 1970s only 8,000 families—a small fraction of the expected total—had been settled along the newly constructed 2,200-kilometer Transamazon Highway. Moreover, cultivation of annual crops on the relatively infertile soils of the area proved to be much more challenging than originally thought. Settlers often discovered that reasonable yields of rice, maize, and other annual crops could be maintained for only a few years after clearing the land. Generally poor agricultural results, combined with rampant malaria and social isolation, prompted many pioneers to sell their plots and move away.

The other major thrust of the government's Operation Amazonia was the promotion of private investment in the region through various tax and credit incentives. Results were mixed. At first many entrepreneurs took advantage of these generous incentives to establish large cattle ranches, mainly in southeastern Pará and northern Mato Grosso. The difficulties they encountered in establishing and maintaining pasture under Amazonian conditions, and their remoteness from the larger urban markets, soon caused many to sell or abandon their ranches. The legacy of these projects was decidedly negative: extensive deforestation and the exacerbation of social tensions over land rights. Although Operation Amazonia gave very little direct attention to rivers, floodplains, and their resources, the deforestation of headwater areas in Mato Grosso, Goiás, Pará, Rondônia, and Acre has had a huge effect on aquatic ecosystems. The most notable alteration is the increased turbidity of rivers, attributable to stepped-up erosion from deforested areas. For example, the waters of the clearwater Rio Ji-paraná (Rio Machado) in Rondônia were rendered milky in color after construction and colonization along the Cuiabá–Porto Velho Highway, which cut across its headwaters.

The Operation Amazonia plans launched by Brazil's military government at first favored the eastern subregion, which now consists mostly of the states of Pará and Amapá. In an attempt to correct this imbalance, Manaus was designated the "development pole" for western Amazonia. In 1967 free trade privileges were extended to the municipality. The special status given to Manaus provided an area of free importation and exportation deep within the Amazon. Free trade designation, along with other fiscal incentives, was intended to nurture an industrial, commercial, and agricultural center in the interior of Amazonia.

The bulk of the free trade zone of Manaus came to consist of large factories assembling electronic equipment, motorcycles, and other goods, and also the manufacture of textiles. The free trade zone factories of Manaus were never competitive on the world market, however, because of the great isolation of the city and transport costs. Meanwhile, downtown Manaus was transformed into a large shopping mall, where imported items were retailed—especially those items (such as electronic equipment and clothing) subject to high tariffs elsewhere in Brazil. The free trade zone is not a sustainable solution to the socioeconomic problems of western Amazonia, since its survival is totally dependent on the continuation of special privileges. Because of Brazil's difficult balance-of-payments situation in the 1990s, the federal government finds itself under pressure to restrict the free trade status of Manaus.

The free trade zone of Manaus had multiple effects on the ecology of the western Amazon, though these have never been studied in detail. The most obvious effect

Manaus, a free trade zone. *The Federal Receipts skyscraper in Manaus has become the economic symbol of the central Amazon. The free trade zone was artificially supported and has done little to promote long-term economic development of the region.*

was the rise in population in the municipality of Manaus. In 1970 there were about three hundred thousand inhabitants, but by the mid 1980s more than a million people lived in the city. The slums of Manaus swelled.

Many people migrated from the interior in the 1960s and 1970s to settle in Manaus and seek employment in the factories and in support activities, such as construction. Because many of these migrants came from rural areas along the western waterways, the boom in Manaus might initially have prompted a population decline in the Amazonian floodplains. By the 1980s, however, the rural population also began to grow—probably because of lower infant mortality attributable to antibiotics and other medicines that became widely available.

Overall, the free trade zone offered no real stimulus or incentive to promote the sustainable use of riverine and floodplain resources. In retrospect, the whole enterprise can be seen as part of the boom-and-bust extractive economy that has dominated the Amazon since the arrival of Europeans. As discussed later in this chapter, the construction boom in Manaus fostered heavy demand for timber, and the floodplain forests were hit hard because of easy transportation. Within a couple of decades most of the valuable floodplain timber within a thousand kilometers of Manaus was gone. Similarly, the regional fisheries were called upon to support the growing population of Manaus, but almost nothing was done in the way of management to ensure sustainability. Floodplain agriculture for food crops remained in a vestigial stage, and thus the inhabitants of Manaus came to depend increasingly on imports.

LOGGING THE FLOODED FOREST

The indigenous peoples probably had only a modest effect on the timber trees of the Amazon floodplain, as the wood in their dwellings consisted mostly of poles used for frames to which either thatch or mud was attached. Of greater impact would have been the felling of large trees to make canoes. Most Indian canoes, however, appear to have been constructed from upland rainforest species, though this matter has never been studied in detail.

Amazon logging on a commercial scale began in the river channels rather than on the floodplains. Under natural conditions, trees on river levees grow right up to the channel's edge. The soft alluvial soils are often undercut during floods, and thus the rivers themselves ensured an annually replenished supply of useful wood in the era before commercial logging. Trees, shrubs, and everything else attached would flow by for the taking. The Portuguese name Rio Madeira, which means "wood river," aptly reflects this phenomenon. During the nineteenth century the quantity of large trees that was brought downriver each year by the floods was sufficient to support small sawmills in Itacoatiara on the middle Rio Amazonas. Mahogany and tropical cedar were especially welcome. These valuable species used to grow on the higher parts of the floodplain as well as the uplands, but today they are very rare except in protected areas along rivers, such as in the Manu National Park in Peru.

One sad reminder of floodplain deforestation is the dwindling amount of wood that Amazonian rivers naturally carry downstream. The waterways bestow fewer and fewer of these gifts every year. As the last levees are deforested, so too will the tree-falls and the mighty trunks and stems that once moved with the torrent disappear.

Log rafts.
With so much of the floodplain of the middle Amazon and its tributaries already deforested, logging operations are now targeting the tidal forests of the estuary.

The refuse of deforestation. *Before deforestation became so advanced, riverbank cave-ins led to an annual supply of huge trees that floated downstream. The flotsam seen today along the middle Amazon River is, however, a sad sign of deforestation.*

The introduction of steamships into the Amazon in the 1860s was the first major economic activity that led to the relatively large-scale logging of floodplain trees. Floodplain species ideal for firewood, such as the smooth-trunked mulatto-wood, were cut in large quantities. By the time diesel replaced steam engines, mulatto-wood trees were rare in many areas. Today only scattered and relatively small individuals of this handsome tree remain along much of the Amazon River.

Large-scale logging in the Amazon began on the floodplains because of relatively easy transportation. With the explosive growth of Manaus and other cities beginning in the 1960s, floodplain logging expanded rapidly to meet construction demands. An economic miniboom ensued, based largely on floodplain logging. In the 1970s huge timber rafts derived from floodplain trees were common spectacles on Amazonian rivers. By about 1980, however, most of the floodplain of the Amazon River had been logged out, at least for the precious species.

The resource has declined so much that now floodplain logging is taking a toll even in presumably protected areas—notably, the Mamirauá Ecological Reserve at the confluence of the Amazon and Japurá rivers. Mamirauá represents one of the few large tracts of natural floodplain ecosystem left along the Amazon River. Considering the history of floodplain logging, it is a shame that this activity is allowed on the only reserve anywhere along the Amazon River in Brazil. The Mamirauá Ecological Reserve belongs to the state of Amazonas, but community-based management by the local peoples rather than strong outside police action is being experimented with as an administrative tool.

Trees are usually cut during the low-water period and then carefully floated out of the forest during the floods. In the estuary, which is subject to tidal rather than seasonal flooding, logs are often placed on skids or rolled into small canals. Each is then pulled out of the forests manually by a dozen or more men. Heavy species such as the piranha-tree and jacareúba-tree sink, and so they must be tied to lighter woods to ensure flotation—which then has a spin-off impact. Most of the large Spruce's rubber trees, a relative of the species of latex fame and an important source of seeds for many animals, were destroyed to make rafts to float heavier woods.

Now little commercial logging takes place along most of the Amazon River. Quite simply, the middle stretch of the floodplain has been logged out. The estuary and western Amazon are the only regions that still draw commercial attention.

Virola, a species valued for cabinetry and plywood, appears to be the most heavily exploited species. The estuarine logging industry (annual sales of about a hundred million U.S. dollars) is based largely on virola. Far to the west, in the Mamirauá Ecological Reserve, one species after another becomes the prime target for cutting, as overexploitation takes its toll. Between the estuary and the western Amazon, the kapok tree—perhaps the most majestic of all floodplain species and once an emergent from the upper canopy that marked the forest's skyline—almost vanished between 1975 and 1985. Loss of the magnificent kapok is perhaps the most poignant reminder of the unsustainability of recent levels of exploitation of floodplain forest.

Meanwhile, no serious reforestation or planting of potentially valuable floodplain timber species has been attempted. Furthermore, no broad study has ever been made of floodplain logging along the inland rivers of Amazonia, though some recent ecological surveys have been conducted in the Mamirauá Ecological Reserve and in the estuary. Domingo Macedo of the Goeldi Museum and Anthony Anderson of the Ford Foundation have calculated that more than 90 percent of the harvestable virola in a particular swamp forest of the estuary was removed in a five-year period.

Drowned forests. *The destruction of huge expanses of rainforest by artificial reservoirs became an issue in the 1980s and 1990s. The environmental effects of the Balbina dam, built on the Rio Uatumã to supply Manaus with hydroelectricity, were especially severe.*

REWARDS AND RISKS OF AMAZONIAN HYDROPOWER

Within ten years of the declaration of Operation Amazonia, the Brazilian government had decided to move away from its initial focus on directed settlement, industrialization, and livestock development. In a major policy shift the government launched the Polamazônia program, which promoted large export-oriented projects in the region. Under this new program, the role of the public sector was to provide basic infrastructure and special tax breaks, while private interests provided equity capital.

Along with widespread disillusionment with the Transamazon Highway settlement scheme, a major factor in explaining this change in policy was the oil crises of 1973 and 1979. The huge price increases hit Brazil, a major importer of oil, particularly hard. This internal shock amplified the country's need for foreign exchange to pay for oil imports as well as to service the country's rapidly mounting external debt. Government officials concluded that exports of Amazon timber, minerals, and agricultural products could help generate the hard currency. The government also planned to harness the region's huge hydroelectric potential as a way of reducing dependence on imported oil.

The cornerstone of the government's "Big Projects" policy was the development of the mineral-rich Carajás region, located about 550 kilometers south of Belém in the state of Pará. To attract private investment, the government offered generous tax holidays, loan guarantees, and abundant energy at subsidized prices from the nearby Tucuruí hydroelectric facility on the Rio Tocantins. The first major project in the Carajás region set out to exploit the huge reserves of high-grade iron ore. This five-billion-dollar project commenced in the mid 1980s. The money supported an 890-kilometer electric railroad, deepwater port facilities in the neighboring state of Maranhão, and urban infrastructure—in addition to mine site development. Despite the colossal nature of the Carajás Project, it has done very little to alleviate poverty in the Amazon.

Of all the big projects, dams have been the most controversial. The Amazon Basin is the largest river system in the world, and water is one of its great energy resources, but the hydroelectric potential of this region is not proportional to its size. The three largest rivers—the Amazon, Rio Negro, and Rio Madeira—flow mostly through low-lying terrain with relatively broad valleys. Stream grade is minimal. The Amazon River drops only fifty-five meters between the Peruvian border and the Atlantic.

One of the more quixotic schemes for the development of the Amazon region was proposed in the 1960s by Robert Panero and Herman Kahn of the U.S.-based Hudson Institute. Panero and Kahn recommended that seven dams be constructed in strategic places around the Amazon Basin, with the objective of creating five South American "great lakes." The promoters' main justification for the scheme was that these lakes would spur regional economic integration by connecting Brazil, Venezuela, Colombia, Peru, Paraguay, and Argentina by inland waterways. More specifically, they argued that trade would be stimulated between the industrial complexes of Buenos Aires and São Paulo and the producers of raw materials located to the north and west of these centers. They also suggested that foreign investment would be encouraged, and that large-scale projects in the areas of energy, forestry, fishing, petroleum, and mining would become economically viable.

The most ambitious and costly component of the "great lakes" initiative was to be the construction of a dam thirty to fifty kilometers long across the lower Amazon River in the vicinity of Monte Alegre. According to Panero, such a dam would create a vast inland sea extending upstream a thousand kilometers beyond Manaus to the town of Tefé. The primary benefits of this undertaking were to be improved conditions for shipping, generation of electricity, and the opening up of new lands for development. The hydroelectric potential of this single dam was estimated to be about a fourth of the installed capacity in the entire United States at the time. For people downstream from the dam, the chief benefit was to be flood control and the opportunity to farm the (former) floodplain year round.

The "great lakes" initiative (thankfully) never got off the drawing boards. If it had gone ahead, it surely would have been one of the greatest human-induced environmental disasters in history. Through drowning or desiccation, it would have destroyed much if not most of the biodiversity associated with Amazonian floodplains. Human use of the floodplain would also have been impaired. Extensive areas of fertile farmland upstream of the dam would have been submerged forever, and the floodplain downstream would have been robbed of the gift of annual fertilizer—the Amazon's silt load. Virtually the only lasting impact of this episode was to reinforce the long-held view of Brazilian nationalists that foreigners really were intent on "internationalizing" the Amazon.

The three main regions of hydroelectric potential in the Amazon Basin are the upland areas to the north and south of the main river and the western Andean foothills. The Brazilian upland areas have a much greater hydroelectric potential than do Amazonian headwaters in the Andean region. The Andean rivers have high sediment loads and thus, as would be the case with the Amazon River, dam reservoirs would rapidly be rendered useless as they filled with silt. In contrast, the upland areas to the north and south of the Amazon River are drained by clearwater or blackwater rivers, which carry very low sediment loads. The largest potential may be south of the Amazon River and east of the Rio Madeira. Rivers such as the Tocantins, Xingu, and Tapajós extend for more than a thousand kilometers; they have low sediment loads and enough stream grade to support numerous dams.

At present Brazil is the third or fourth largest producer of hydroelectric power in the world. And Brazil has developed only about 4 percent of its hydropower potential, compared with the 24 percent of capacity already tapped by the United States. These numbers can be misleading, however. "Potential" hydropower is usually calculated from a scenario in which every river is dammed at all possible sites. An environmental cost is usually not factored into the "potential" hydropower equation. Thus these numbers should be viewed with great care. In the case of the United States, for example, far more than a fourth of hydropower potential has already been reached if unsatisfactory environmental and social costs are to be avoided.

Brazilian hydroelectric potential in the Amazon is calculated by Electrobrás and its subsidiary Electronorte. Electrobrás is a government monopoly that controls virtually all major power resources in the country. The most widely quoted estimates of hydroelectric potential and megawatt-per-hour costs keyed to particular sites is drawn from the so-called 2010 Plan. This plan, proposed by Electrobrás in the early 1980s, entails an ambitious schedule for dam construction to the year stated in its title.

The Amazon, without a doubt, has one of the largest untapped hydroelectric potentials in the world. Dam building in the Amazon, especially on the scale proposed by various Brazilian governments, would also without a doubt lead to a much greater destruction of biodiversity than almost anywhere else on Earth. Yet the Amazon region, with its rich mineral resources, explosive growth in urban areas, and depressing levels of poverty, cannot be expected to develop and move beyond peasant economies without additional sources of energy. One cannot expect Amazon economies to be fueled mostly by biogas, solar panels, wood, and windmills.

Conservationists and human rights groups usually oppose hydroelectric projects for their negative impacts on the environment and native peoples. But until economically attractive alternatives become possible, dam proposals will continue to draw support. The natural gas and oil discoveries in the Brazilian Amazon that are now being announced could perhaps shift the energy focus back to thermoelectric power, and thus weaken the drive to pursue the large dams.

Tucuruí, the main source of electric energy for both Belém and the Carajás mining and metallurgical complex, was the first major power project in the Amazon region. This 1,200-meter-long dam spanning the Rio Tocantins commenced operations in 1984. Currently, Tucuruí has an installed hydroelectric capacity of 4,000 megawatts, and there are plans to almost double this capacity by the end of the 1990s. Two older and far smaller hydroelectric projects, Coaracy Nunes (40 MW) and Curuá-Una (30 MW), were completed in the mid 1970s to supply the local markets of Macapá and Santarém, respectively. The Balbina and Samuel dams, both in the 200–250 megawatt range, were completed in the late 1980s to provide energy to the cities of Manaus and Porto Velho, respectively.

In terms of energy supply Tucuruí has had a favorable impact on the economy of Pará. The large city of Belém now has a reliable source of energy, whereas before the dam was built daily power blackouts were common. It can also be argued that the rapid growth of the region's other major population centers has made an increased supply of energy an absolute necessity. The financial viability of energy-intensive metallurgical plants has also improved, though some industries have been granted relatively low energy rates by the Brazilian government in order to spur development of regional mineral production.

But what of the environmental effects of the Tucuruí project? The environmental studies that have, to date, been carried out on the Rio Tocantins ecosystem are very superficial in comparison to the great ecological changes that dam construction anywhere inevitably causes. We still do not know, for example, to what extent the aquatic ecosystems have been affected, though some possibilities are discussed in chapter 5. In contrast, Phillip Fearnside of the National Institute of Amazonian Research judges that the Balbina Dam near Manaus is one of the greatest public works fiascoes ever undertaken in the Amazon. And he is not alone in his assessment. Although Balbina produces only about 5 percent of the energy that Tucuruí does, it drowned a rainforest area of about the same size—approximately 2,500 square kilometers.

The construction of dams in Amazonia will surely continue to draw criticism for the damage they do to both forest and water resources, and for their disruption of riverine communities. Over the current decade, six additional hydroelectric projects are planned for the Amazon region. If implemented, they would add 3,336 square kilometers to the 5,437 square kilometers already artificially flooded. At present the environmental problems associated with dams are barely being studied in the Amazon.

GOLD FEVER COMES TO THE AMAZON

Promising discoveries of natural gas have been made in the western Amazon region, and oil is already exploited in Ecuador and Peru. But the earth's crust in the Amazon Basin seems to be far more richly endowed with minerals than with hydrocarbons. Important mineral assets include bauxite, high-grade iron ore, manganese, and (above all) gold.

In the last decade gold has been the single most valuable resource exported from the Amazon Basin. Annual estimates of Amazon gold revenues range between one and three billion dollars. Despite impressive growth in the mineral sector, little scientific research has been done on the environmental effects of mining in Amazonia. At

Gold fever.
The Amazonian "truth," as ironically suggested by the name of this processing shop, became gold in the 1980s. Wildcat gold mining led to sediment and mercury pollution, and these problems continue today.

least one such effect is far from subtle—mercury pollution is known to be both widespread and severe.

For more than a decade the Amazon Basin has witnessed a gold rush that now embraces all of Brazil's states and territories within this region. Most of the Amazon gold miners operate independently in the sense that they are not linked to corporations and, historically, their activities have not been recognized as legal. In Brazil this type of wildcat mining is called *garimpo* or *garimpagem,* and the miner is referred to as a *garimpeiro.* Wildcat miners also exploit cassiterite, the ore from which tin is made.

Brazil has a long history of wildcat mining. *Garimpeiros* have pursued fortunes in the Amazon since at least the seventeenth century. Estimates of the numbers of wildcat miners present in Amazonia today range from 300,000 to 500,000. Perhaps as many as a million people are directly or indirectly involved in gold mining. These numbers, combined with the small scale and requisite furtiveness of the clandestine operations, make the mercury problem especially intractable.

The main Amazon gold rush dates from only about 1979. That was the beginning of the phenomenal migration to the rivers and rainforests in search of the precious ore. In 1979 the international price of gold reached $700 per troy ounce, and it has rarely dropped below $300 since then. (There are twelve ounces in a troy pound, the measure used in the United States for precious ores.) Coincident with the sharp rise in gold prices was a severe drought that struck northeastern Brazil beginning in 1979. The failure of the Brazilian government to provide drought relief and to develop alternative programs for the unemployed spurred a massive migration to the Amazon Basin in search of new lands and gold. In some ways this pattern parallels the migration of northeastern Brazilians to the Amazon Basin in the latter part of the nineteenth century, when rubber was the most valuable product exploited. Rubber tappers were confined mostly to the Amazon lowlands, whereas gold miners are concentrated in the middle and upper reaches of tributaries draining the Brazilian and Guiana shields.

To understand the forces that create and sustain Amazon gold fever, one must begin with the Brazilian economy. According to official estimates, from 1977 to 1983 gold extraction in the Amazon jumped from 1.6 to 47.5 tons. A national recession, coupled with the drought in the Northeast, pushed many desperate men to migrate to the Amazon gold fields. During this same period, Brazil's central bank purchased large amounts of gold in order to boost its reserves that, theoretically, would help the country deal with its financial crisis. One must also recognize, however, that it was the gold extracted by wildcat miners in the Amazon that made the central bank's policy possible.

In 1983 gold production peaked at Serra Pelada in the Carajás region, which is in the southern part of the state of Pará. Serra Pelada contains the most famous wildcat field in the Amazon. Ten tons of gold were extracted from that site alone at its peak in 1983. Wildcat mining became increasingly mechanized around that time, and thus total output of gold in the Amazon increased despite the decline of Serra Pelada. At first the Rio Tapajós and Rio Madeira valleys, in the states of Pará and Rondônia, respectively, were the focus of the rush. By 1990, however, more than 50,000 hopeful miners were reported to have migrated to the northern state of Roraima, the newest frontier. As of 1990 Pará was the largest gold producer in Brazil, followed by Mato Grosso. Present trends suggest sustained growth in gold production in these two states, and in Roraima.

Wildcat miners account for an estimated 85 percent of all gold extracted in Brazil. By 1987 Brazil was producing more than a hundred tons of gold annually, making the country the world's fourth largest producer. According to the National Department of Mineral Research, 95 percent or more of all Brazilian gold production comes from the Amazon region. As little as a third or even a fourth of all gold extracted in the Amazon, however, is sold through official channels. For example, Brazil's official records of cumulative gold production from 1977 to 1986 report only 167 tons of gold from all sources. The National Department of Mineral Research, however, calculates that 434 tons is a more reliable figure of actual production during the same ten-year span.

Beginning in 1979 the international price of gold increased rapidly, as investors turned to it in reaction to the second oil shock and the rising interest rates propelled by the United States' attempt to curb inflation. Coupled with an uncontrolled wildcat mining sector in Brazil, this set the stage for the contraband market that came to dominate the Amazon gold business in the 1980s. The Brazilian government lost hundreds of millions of dollars in potential revenue because of its inability to police the gold market and implement taxation. Surely some of this unrealized revenue could have been used to tackle the problem of mercury pollution.

Amazon wildcat gold mining is mostly placer rather than the hard-rock variety. Techniques for separating the precious ore from lighter sediments have passed through three distinct technological phases. In the pre-1978 period Amazon gold mining was characterized by lone or small groups of prospectors, manually working deposits that were usually far removed from urban centers. Rapid mechanization began in 1978, the most important innovation being the use of suction pumps with hoses operated from rafts and barges. Finally, heavy machinery, such as tractors and steam shovels, was introduced to work deposits near the many new roads and other access points. Amazon wildcat gold mining now involves the use of several thousand barges, some 20 helicopters, 750 small planes, and perhaps 10,000 boats or motorized canoes. In general, greater mechanization has led to increased pollution.

THE TURN TO FISHING AND CATTLE RAISING

Traditional riverine communities were for the most part unaffected by Operation Amazonia and later policies, which focused almost exclusively on the occupation and development of lands above the floodplains—the terra firme. The floodplain was not, however, completely forgotten in the official development plans. For example, the first five-year development plan for Amazonia, covering the 1967–1971 period, declared that the floodplain was the preferred zone for cultivating annual crops. Recognizing that Amazon soils beyond the floodplain are generally poor, the plan recommended that agriculture on the terra firme center on perennial crops and extensive livestock raising. The plan called upon the regional development agency (SUDAM) to promote the rational use of the floodplain through studies and special development projects.

Despite the official rhetoric extolling the agricultural potential of the Amazon floodplain—echoed in virtually all regional development plans that followed—little public investment has been made or is forthcoming. The development of the floodplains seems never to have achieved the political prominence given to the settlement of the terra firme. It is revealing that a Brazilian national plan called PROVÁRZEAS, which was launched in 1981 for the irrigation and agricultural development of the floodplain, had little impact on Amazonia. That program was, in fact, largely directed to the prosperous South, where both labor and capital were more plentiful.

With the collapse of the jute economy, many floodplain inhabitants have been obliged to develop alternative sources of cash income. The most important have been commercial fishing, livestock raising, and the cultivation of fresh fruits and vegetables for sale to local markets. The growth of these enterprises has to a large extent been driven by the rapid growth of the region's major riverfront cities.

Although supporting data are fragmentary, it would appear that many former jute farmers have turned to commercial fishing. The relative economic importance of agriculture and fishing has thus reversed. Fishing, which was once mainly a subsistence activity, has now become the principal source of cash income. Agriculture, once a major source of cash income, has become mainly a subsistence activity.

The traditional domain of floodplain fishing has been the floodplain lakes. In the past, local populations had these lakes practically to themselves. In recent years, however, floodplain dwellers who look to the lakes for their livelihoods have come into direct competition with urban-based fishermen operating out of Belém, Santarém, and Manaus. The growth in the numbers and reach of these urban fleets reflects both expanding markets and major improvements in fishing technology.

In addition to commercial fishing, riverine populations have also become much more involved with livestock raising in recent years. Animal husbandry is not a new economic activity on the Amazon floodplain. Cattle were brought to the floodplain during the colonial era, but most grazing took place on natural grasslands, particularly in Roraima and Marajó. Many floodplain peasants have traditionally kept a few head of cattle as insurance against downturns in their primary modes of livelihood. Given Brazil's history of inflation, the relatively low interest rates paid on savings accounts, and the collapse of jute prices, peasants have begun to believe that cattle are a better investment than monetary deposits. Then too, the same urban growth that stimulated commercial fishing has also swelled demand for beef and dairy products. Livestock raising—a topic we shall revisit in chapter 8—is now the dominant form of land use on the Amazon River floodplain between Manaus and the Rio Xingu, a distance of nearly two thousand kilometers.

CHAPTER FOUR

A WEALTH & WASTE OF WILDLIFE

The 2 to 3 percent of the Amazon Basin that is floodplain is home to a great diversity of animal species. Most of the non-aquatic vertebrate animals found on floodplains also inhabit the uplands. However, the floodplain is ecologically more important to arboreal and terrestrial animals than its size alone suggests because its flooded forests supply fruits, seeds, and other foods at times when they are scarce in the uplands. There is no evidence that any vertebrate animal on the Amazon River floodplain has yet been driven to extinction, but the populations of many species have plummeted along the middle and lower stretches of the floodplain because of deforestation and hunting pressure.

Most of the Amazon's legendary biodiversity is not, however, expressed in the vertebrates. Biodiversity owes largely to the insects—notably, to tens of thousands of species that have not yet been scientifically described, and whose habitats may be extremely limited in range. Scientific collections of animals of any kind are expensive to make and even more costly to maintain. North American and European museums do maintain huge collections of most Amazon animal groups. The museums of South American countries, and especially those of Brazil, have recently made great strides in attaining large scientific collections, but a worldwide effort is essential if most of this material is to be properly classified so that biodiversity analyses can be made on a large scale.

Owing to limited financial resources, one of the great policy issues flowing from the biodiversity debates today is to what extent more collections need to be made before or while specific actions are taken to preserve biodiversity. Many, if not most, areas in the Amazon, and this includes the floodplains, have not been satisfactorily inventoried for animal diversity. And consider: at present we do not know the life cycle of even one insect species that inhabits the flooded forest canopy. Thus, even if we learn a great deal more about what species are there, we will still not know which plants and animals those species need to survive.

Some biologists feel that since rainforest diversity is now threatened, every effort should be made to collect as many species as possible—even if museum funds are insufficient to study the specimens. According to this view, somewhere down the road future generations will be able to determine what the animal biodiversity was like before deforestation took place. This information would also be essential for evolutionary studies, and especially for determining the geographical origins of species. An alternative view that is gaining support urges that limited funding should be channeled not into new collections but into interpreting the collections already held. Existing collections could provide the basis for extrapolating the kind of distributional and ecological information that would help in determining where parks and reserves might best be located to preserve biodiversity. This chapter cannot resolve the debate, but it should help the reader get a better idea of the nature of animal biodiversity found in Amazonian rivers and floodplains and its use by humans. More attention is given to vertebrate animals for the simple reason that much more is known about them than about invertebrates.

THE PLIGHT OF BIRDS

Birds are the most diverse group of vertebrates found in the canopy of the flooded forest. The Amazon Basin as a whole, in all its habitats, is home to at least 950 bird species—a tenth of the world's entire avian biodiversity. Almost half are endemic; they are found nowhere else. The floodplains alone are home to perhaps 350 to 400 bird species.

Rivers and their floodplains play an extraordinary role in the distribution of Amazon birds. A large river usually marks the limit of a bird's distribution within any particular region of the Amazon Basin. Relatively few Amazon bird species are limited to floodplain habitats. Many upland species, however, spend part of their time in floodplain forests to roost, breed, and feed. During the dry season some upland forest birds seek proximity to water.

Macaws and parrots are among the most striking examples of birds that rely on floodplains, and especially those of the river islands. Many macaw and parrot species roost in upland forests or swamps away from the floodplains. During the early hours, however, they return to the floodplains and river islands to feed, flying back to roosting sites before sunset. Some of the toucans and various songbirds also move diurnally to and from the floodplains. Where floodplains and river islands have been heavily deforested, such as along much of the middle Amazon River, macaws and some of the larger parrots are now seldom if at all seen. What is unclear, but in urgent need of study, is to what extent the disappearance of floodplain forests has diminished macaw and other bird populations that depend on both upland and floodplain habitat.

Unlike macaws and parrots, the hoatzin and umbrellabird are not popular pets. But they are valuable to humans in another sense: they may have a lot to tell us about the evolution of floodplain birds. The hoatzin is the only living member of its family.

Its strange-looking juveniles possess a claw at the bend of each wing to help them scamper back up to the nest when they drop into the water to escape predators. The hoatzin is also unique because it has a double crop. A double crop gives this bird a physiological affinity to ruminant mammals; the hoatzin is able to maintain a community of symbiotic microorganisms that help it digest leaves.

Both the hoatzin and the umbrellabird almost certainly evolved within the Amazonian floodplains. The Amazonian umbrellabird does not tolerate floodplain deforestation, and it has disappeared along most of the Amazon River where it had been

Victims of deforestation. *Because blue-and-yellow macaws require tall floodplain forest, deforestation along the middle Amazon has taken a severe toll on the populations of these and many other large parrots.*

The strangest bird of the flooded forest.
The hoatzin is the world's only leaf-eating bird, and its young have curiously "clawed" wings. When hoatzins disappear from a floodplain area, it is an ominous sign that many animals are at risk.

reported by naturalists in the last century. It is now confined mostly to some of the tributary floodplain forests. The hoatzin tolerates some habitat disturbance, but its disappearance from an area is an ominous sign that deforestation has reached the point where forest bird populations in general have collapsed. Although hoatzins feed on herbaceous plants such as arum—a large free-standing plant in the philodendron family—they depend on floodplain forests for nesting sites.

Most of the large birds of the Amazon floodplain depend on forest for nesting sites. These include many species of raptors, nearly all herons, egrets and ibises, ducks, toucans, owls, and many others. Many of the egrets, herons, and storks also nest in arum patches. The philodendron-like arum is a herbaceous plant, but because of a sturdy stem and communal growing habits an arum can reach seven to ten meters tall. An arum patch thus functions, for nesting birds at least, as a miniforest. Arum grows in low-lying places that, under natural conditions, seldom dry out. When deforestation and repeated burning take place nearby, however, arum patches may wither during the low-water period. Individual arum plants at the edge of the community then topple and die, owing both to desiccation and to trouncing by introduced water buffalo.

The destruction of nesting sites is the single most important factor reducing the biodiversity and abundance of large birds along the Amazon River. Hunting and egg collecting for food are also taking a heavy toll on egrets, herons, ibises, and ducks. If present deforestation and hunting trends continue for another decade or two, the wading birds of the Amazon River, which were once extremely numerous, will probably become rare.

The increase in floodplain meadows as a result of deforestation may have initially expanded feeding habitat for some bird species. Many of the finches, blackbirds, anis, and tyrant flycatchers feed on the seeds or invertebrates found in floating meadows and floodplain grasslands. The combination of cattle, buffalo, goats, sheep, and pigs, however, results in heavy grazing that leads to the annual destruction of herbaceous communities when livestock are relocated to the floodplains as the floods wane.

The large-scale introduction of livestock onto Amazonian floodplains has offered great opportunities to the cattle egret, a species native to the Old World. The cattle egret first colonized the Amazon Basin in the 1960s by way of the island of Marajó.

Wading birds that rely on forest.
Although the great egret and many other large wading birds forage in shallow waters and floating meadows, they depend on trees or arum patches for nesting.

This bird reached the New World on its own, but has rapidly dispersed because of the presence of livestock. It is now the most common wading bird of the Amazon floodplain. The species gets its name from its habit of accompanying and perching on livestock. Cattle egrets feed on the invertebrates scared up or turned up when the large herbivores trundle through grasslands or floating meadows. Cattle egrets now dominate the few remaining nesting sites traditionally used by wading birds. A combination of reduced nesting sites and their dominance by cattle egrets suggests a diminished role for many Amazon floodplain wading birds. A possible solution to this would be to encourage the collection of cattle egret eggs by local peoples, while discouraging the taking of eggs of the indigenous species.

The direct importance of Amazonian floodplain birds for human economies is at present minimal. There are small, and technically illegal, markets for feathers, eggs, and meat. Hunting is principally for subsistence, though wanton destruction of large birds, under the guise of sport, is common. The Neotropic cormorant is probably the bird most regularly killed for food. Its populations have been reduced greatly in the past decade because of hunting. Local hunting folklore has created an almost must-kill attitude toward large birds, such as the buff-necked ibis. If seen they are shot at. The same goes for muscovy ducks and maguari storks. Muscovy ducks have been domesticated, so they are unlikely to vanish from the Amazonian floodplains. Nevertheless, the large-scale decline of wild populations has reduced the potential gene pool that could be tapped for improving these food animals.

Cage birds are very popular pets among Amazon residents, and little has been done to control the illegal exploitation of wild species. With the explosion of urban populations in the last two decades, the demand for cage birds has greatly increased. Although macaws are mostly taken from nesting sites in palm swamps or upland forests, the floodplains—owing to their accessibility during the flood season—are probably the hardest hit habitats for the cage-bird trade as a whole. Especially popular species from the floodplains include the beautiful and sonorous troupial (a type of oriole),

Target of the cage-bird trade. *The troupial is one of the most beautiful songbirds of the Amazon floodplain.*

canary finches, parrots, parakeets, and parrotlets. Deforestation and the cage-bird trade have greatly reduced the numbers of troupials on the Amazon floodplain. Troupials lay their eggs in the hanging basketlike nests of caciques or oropendolas. Since these nests are easy to find, the troupials are an easy target. There have been no serious attempts in the Amazon to breed cage birds. Some of the macaws and parrots have been bred elsewhere, such as in Florida, but Amazonians, despite their fondness for cage birds, have shown almost no interest in aviculture other than chicken farming.

MAMMALS BIG AND SMALL

The Amazonian lowlands support about 200 mammal species. Half of these inhabit floodplains, though most not exclusively so. The floodplain mammalian diversity is largely found in, and depends on, trees. No major Amazon mammal group, with perhaps the exception of rabbits, is missing from floodplains. Bats and rodents are the most diverse floodplain mammals, but also the least known taxonomically and ecologically. Bats, rodents, and monkeys are represented by species endemic to the Amazon, including some found only or mostly on Amazonian floodplains. The tucuxi dolphin and Amazonian manatee are also endemic to Amazonian rivers.

No convincing hypothesis has yet been proposed to explain why some mammal species are confined to floodplains. For example, the world's smallest monkey, the pygmy marmoset, feeds largely on the sap of floodplain trees. Another primate group restricted to Amazon floodplains is the uacari monkeys. These monkeys are fortunate—thus far at least—that the floodplains of the western Amazon (their home range) have suffered much less deforestation than those of the eastern part of the basin. It is highly unlikely that uacari monkeys would survive in floodplains as deforested as those of the middle to lower Amazon.

New World monkeys, unlike their counterparts in the Eastern Hemisphere, cannot easily digest leaves. They depend on fruit—which makes the fruit-rich floodplains a favored habitat. One exception is the group of howler monkeys. Bacteria in the hindgut allow howlers to include leaves in their diet. Capuchin as well as howler monkey populations along the middle and lower Amazon have, however, been decimated by a combination of deforestation and hunting. Both monkeys used to be common on river islands as well, but just how they got there in the first place is unclear. It is also not known whether these island populations were isolated long enough to evolve significant genetic differences from upland populations.

The only floodplain primate species that has been able to adapt to heavy deforestation is the squirrel monkey. This species (which also thrives in the uplands) can make do with even very small patches of forest, and it is little hunted for food. Though many isolated populations of squirrel monkeys can be seen on the Amazon floodplain, it is not at all clear whether these can survive in the long term, as isolation restricts their genetic diversity and makes them vulnerable to local extinction by disease. Moreover, even the remaining patches of forest are threatened by livestock ranching.

In some floodplain forests, sloths may account for most of the mammalian biomass. With stomachs divided into many compartments, sloths enjoy the highest adaptations for leaf eating of any mammal in the flooded forest. Surprisingly, sloths are

World's smallest monkey. *The pygmy marmoset lives only in floodplain forests of western Amazonia.*

excellent if slow swimmers. They are known to take to the water to cross even the Amazon itself.

As in the case of birds, many kinds of mammals migrate between upland and floodplain during the course of a year. Floodplains produce an abundance of fruit and seeds during the rainy season. Howlers, capuchins, titis, and other monkeys move to the floodplains then to feed. Some of the frugivorous bats probably do the same. Ground-dwelling species, such as peccaries and tapirs, move onto the floodplains several months later. As the water recedes they search for fallen fruits and seeds. Both peccaries and tapirs swim to islands that emerge as the floods ebb. Large predators, such as the jaguar and ocelot, follow their prey to the floodplains and even swim out to the islands to find them. Seasonal use by large mammals of emerged islands and other vulnerable habitats has, however, largely disappeared from the Amazon River floodplain because of human colonization and hunting. But peccaries, tapirs, and their predators can still be observed in some of the tributaries, such as the Rio Negro. Floodplain ranchers claim that jaguars kill livestock, but this loss is greatly exaggerated.

The Amazon Basin has few large grazers. The two main exceptions are the manatee and capybara. The manatee, a sirenian, is South America's largest animal; the capybara is the world's largest rodent. Both feed mainly on herbaceous plants of the floodplains—notably, the floating meadows that appear and spread with the floods. Both acquire huge amounts of body fat during the floods, on which they survive the

A primate endemic to the flooded forest.
The white bald uacari does not venture out of the flooded forests of the western Amazon. It thus joins the pygmy marmoset and one other species of uacari monkey as the primates most threatened by floodplain deforestation.

meager fare of the dry season. These two herbivores have been, and continue to be, heavily hunted for food and local sport.

The manatee is threatened now by its very size—which renders it one of the animals on the masculinity-enhancement list of macho Amazon fishermen and hunters. Domestic water buffalo compete with the manatee for the same food, but unlike manatees they trample the shores. Meanwhile, floodplain capybara populations are rapidly losing habitat in which to hide because of the denuding effect of livestock grazing. They are thus being forced into thick arum patches that hunters seldom penetrate. Hunting pressure has caused capybara populations to alter their bathing behavior. Capybaras in times past bathed in the river during the morning and midday hours. This activity is extremely difficult to observe along the Amazon River today.

World's largest rodent. *The semiaquatic capybara and the fully aquatic manatee are the only large native mammals that graze the floating meadows of Amazonian rivers. Hunting and habitat destruction have taken a severe toll on both populations.*

The manatee and capybara have been touted as potential species for domestication. The manatee may surpass three hundred kilograms in weight, but its growth and calving rate is probably too slow for commercial ranching. The capybara, reaching sixty kilograms, appears to be much more promising. Raising the capybara on the floodplains would certainly do much less damage to the environment than cattle and buffalo ranching. The capybara, moreover, breeds easily in captivity. Although some peasants have clandestinely experimented with capybara farming, an overwhelming focus on cattle and buffalo ranching by governments and entrepreneurs seems to be a key reason that interest in capybara ranching has languished. Added to this is the fact that capybara farming is technically illegal.

CAIMANS, TURTLES, AND OTHER REPTILES

At least 300 species of reptiles inhabit the Amazon Basin, of which perhaps half can be found on the floodplains. Although much taxonomic work remains to be done, thus far snakes appear to be the single most diverse reptile group in the Amazon Basin, where more than 175 snake species have been described. Lizards are not far behind. The remaining reptiles include legless lizards, crocodilians, and turtles. Though few in species (and now numbers), turtles traditionally played a prominent role in local cultures and economies.

Many snake and lizard species have relatively restricted distributions on Amazonian floodplains and in adjacent humid areas. In contrast, most turtles are aquatic and widely distributed. The two land tortoises of the Amazon move onto the floodplains during low water, but they rarely cross lakes or channels. Although crocodiles have been successful in most of the major tropical river systems of the world, they are absent from the greatest river system of all. The American and Orinoco crocodiles are found only as far south as Venezuela and Colombia. The only crocodilian kin in the Amazon are four species of caiman, which are more closely related to alligators than to crocodiles.

Both the lizard and snake faunas include species that live amphibious lives, though the majority of floodplain representatives of both groups are probably arboreal. Some of the large lizards of the family Teiidae live mostly in the treetops but feed in the water. Iguanas feed in the trees or on the ground, but not in the water. When danger approaches, however, iguanas readily drop out of the flooded forest canopy and into the water. Unlike many parts of Central and South America, in the Amazon iguanas are rarely eaten. These lizards can, however, be farmed, and some floodplain plants, such as arums, might be cheap sources of food for them.

As might be expected, Amazonian floodplains have a relatively wide variety of aquatic or semiaquatic snakes. Like snakes everywhere, the Amazonian species are carnivores. The world's largest snake, the anaconda, is a common predator and lives mostly in the water. Other floodplain constrictors, such as the boas, take to the trees during the floods. Both main groups of poisonous snakes, the pit vipers (Viperidae) and corals (Elapidae), are well represented on Amazonian floodplains or along their edges. Coral snakes can live in inundated floodplain areas, such as floating meadows, whereas the pit vipers apparently move from ground to trees during the floods. Most

An amphibious teiid lizard. *Teiid lizards live in floodplain trees but feed on mollusks and other small animals found in the water. They are one of many groups that disappear with deforestation.*

snakes in the Amazon belong to the nonpoisonous and nonconstricting family Colubridae. Many colubrids are often mistaken for poisonous snakes, however. Several aquatic colubrids live mostly in floating meadows, where they feed on fishes, frogs, and other organisms. Many of the colubrids, such as the vine snakes, are arboreal and do not survive without forest habitat. Other colubrids prefer the ground, taking to the vegetation only during the floods, but easily and readily moving about in the water.

Snakes are of little direct economic importance since they are seldom eaten. Pit vipers are sometimes blamed for cattle deaths. Anacondas, despite their nasty reputation, present no danger to animals as large as humans and cattle, though they do attack domestic fowl if they can get to them at night. Boa constrictors are occasionally kept as ratters.

The nineteenth-century naturalists who visited the Amazon were greatly impressed with the dense populations of caimans along the rivers and floodplain lakes. Henry Walter Bates described the numbers of Amazon caimans in the 1850s as being as thick as tadpoles in a summer pond in England. By the mid 1950s, alas, the biggest Amazonian caiman, the black caiman, had largely been killed off, and hunters turned to the smaller spectacled caiman. Caiman hunting for both skins and food continues today, though it has long been illegal, at least on the books.

Because of hunting pressure, only in a few isolated areas is it now possible to see large numbers of caimans sunning on a beach or floodplain lakeshore. Of the four crocodilians found in the Amazon Basin, the black caiman is the only species endemic to the Amazon (or nearly so, since it is also found in the Guianas). The black caiman, which may grow to more than four meters in length—the size of an alligator—is also the most threatened species.

In the early part of this century, the international skin trade began to take a toll on Amazonian caimans. Only occasionally were these reptiles killed for food. Eggs were taken as food by peasants, but the harvests were insignificant, especially compared to attacks on turtle nests. By the 1960s caiman populations along Amazonian rivers were so reduced that Brazilian and international conservation interests successfully pushed for a prohibition on hunting and especially exports. The caiman skin market was then driven mostly by demand in Europe and the United States, and so import restrictions also proved to be an important conservation tool. Despite still-decreasing wild populations and export restrictions, no serious attempts have been made in the Amazon to farm caimans. Successful farming experiments have taken place in Colombia and Central America, and parts of the Amazon floodplain would be very appropriate for this activity.

No animals, either for their meat or eggs, are more appreciated in Amazon cuisine than aquatic turtles. Turtles and their eggs are so much talked about and sought after as food, despite the illegal status of such uses, that this passion could be considered a cultural mania.

Considering the size and river complexity of the Amazon Basin, its turtle fauna, with about twenty species, is not particularly rich. The turtle fauna of the Mississippi system, for example, is more diverse. Four side-necked aquatic species (family Pel-

Once populous, now almost gone. *The black caiman was severely overhunted in this century for the skin trade. Only in a few places along the Amazon River can this largest of all Amazonian reptiles still be seen.*

omedusidae) dominate the turtle fauna of Amazonian rivers and floodplain lakes. Before large-scale hunting and egg gathering devastated their populations, Amazon turtles were extremely abundant.

The giant Amazon river turtle, which is the largest freshwater chelonian in the world, is the most endangered species. The highly social nesting behavior of this species is more like that of the giant sea turtles, to which it is not directly related, than to other members of its own family. The giant Amazon river turtle nests socially on beaches during the low-water period. Unfortunately, very little is known about the numbers and exact locations of nesting beaches that existed before the populations of these turtles were greatly reduced. We do know, however, that there were large nesting sites in the Rio Madeira and along the Amazon River in the nineteenth century—but there are none today. The only sites where large numbers of turtles can now congregate safely are in a few protected reserves, such as those of the Rio Xingu and Rio Trombetas in the state of Pará.

Henry Walter Bates noted as early as the 1860s that the Amazon giant river turtle would become exceedingly rare if egg predation by humans continued at the intensity evident to him even then. Bates estimated that 48 million eggs, or the yearly offspring of 400,000 turtles, were being harvested annually by humans. Bates's prediction, sadly, has proved correct. Egg collecting has taken a severe toll on the populations, which have also suffered from the exploitation of adults. Today the giant river turtle is too rare to serve as a major food animal, but it is nevertheless intensively exploited in the illegal game trade—where one specimen can earn for its seller as much money as a legal job might pay over the course of two or three months. Wealthy Amazonians consider this giant reptile a necessity on the menu for birthdays, weddings, and other celebrations.

The smaller side-necked turtles, in contrast, are not social nesters. Their eggs are thus more difficult to find. Solitary nesting is the main reason that the various smaller species have survived in much larger numbers than has the giant river turtle. People who live along the rivers, however, search almost every beach, bank, and floodplain habitat where solitary turtles are known to lay eggs. The destruction of floodplain vegetation is probably the main factor that now makes it so easy to find the dispersed nests of the nonsocial turtles. Fire is also used to clear underbrush in order to reveal nests. The Rio Negro is probably the last large river where turtles have survived in sufficient numbers to be of importance to the local human food supply, though egg collecting is technically illegal.

Turtles of all types are further threatened by floodplain deforestation. Similar to fish, turtles migrate to the flooded forests during the high-water season to feed on fruits and seeds. The giant river turtle, for example, feeds heavily on the seeds of rubber trees.

Since the 1970s several Amazonian countries have conducted small-scale experi-

Seasonal users of flooded forests. *Various species of side-necked turtles migrate to flooded forests to feed on fallen fruits and seeds. They are threatened by hunting, egg gathering, and floodplain deforestation.*

ments with turtle farming. Because of great difficulty in getting these reptiles to breed in captivity, commercial prospects for turtle farming are not yet promising. Artificial water-level changes have been experimented with to induce nesting behavior, but to no avail. As has been the case with many fish species reared artificially, it will probably be necessary to develop hormone injections to induce turtle breeding in captivity. Brazilians have been quite successful in developing hormone injections for fish, and there is no biological reason why chemical counterparts could not be designed for turtles.

Methods to protect nesting sites of the giant Amazon river turtle from natural predators have also been tested. These predators include caimans, catfishes, dolphins, vultures, ibises, and storks. The birds prey upon eggs. Caimans, catfishes, and dolphins attack young turtles when the hatchlings enter waters after emerging from sand nests. Proactive measures to protect eggs and hatchlings from natural predators is intended to increase the chances of the turtles rebuilding their populations. The hoped-for end will remain elusive, however, as long as human predation remains uncontrolled. In the short run this sober assessment suggests that turtle hatcheries might indeed be important for the long-term survival of the giant river turtle.

FROGS OF THE FLOODPLAINS

The Amazon Basin probably has somewhere between 250 and 300 frog species. Biologist Barbara Zimmerman documented 80 frog species within a few square kilometers of upland forest near Manaus. The floodplains are less well known, but a comparable area usually reveals at least 20 or 25 species. About half of the floodplain species are found only there.

One of the most striking facts about Amazon frogs and floodplains is the very poor showing of aquatic species—that is, species that stay mostly in the water. The floodplains of the sediment-rich rivers would seem an ideal place for aquatic frogs. These and other quiet waters with a wealth of insect life suggest frog habitat, but few of these amphibians live in open waters or lay their eggs there.

Worldwide, the family Ranidae accounts for most of the common frogs associated with quiet waters, but this group is barely represented in the Amazon. The poor showing of aquatic frogs in the Amazon could be due either to the failure of ranids to colonize the Amazon or to the presence of an incredible variety of predatory fishes for which the majority of frogs have little defense. Those species that do place their eggs in floodplain waters, other than in isolated pools, often have toxin-producing (and thus distasteful) tadpoles. Most of the floodplain frog fauna live either in the trees or on top of the floating meadows, where they are protected from predatory fishes. Because so few Amazonian floodplain frog species are aquatic, most depend on the forests or floating meadows for their survival.

Frog farming is a growing business in southern Brazil. Northern Brazilians, however, have little appetite for these animals. But recent immigrants, such as those from Paraná, could change this. Whether any of the frog species native to the Amazon would be marketable is unclear. A frog known scientifically as *Leptodactylus pentadactylus* is large enough to be used as food, though care would have to be taken with irritants exuded from its skin. The large Bufo toads, some reaching a kilogram in weight, are too tabooed to be eaten because of their reputedly "ugly" nature and skin toxins. Pipa frogs, which are aquatic species, are captured for the aquarium trade. They breed easily if kept in small ponds. Pipa frogs thus offer an economic opportunity that could complement the farming of ornamental fishes.

HARVESTING SHRIMPS AND MOLLUSKS

The Amazon Basin has a relatively rich freshwater shrimp and crab fauna, with each group represented by at least thirty species. An average floodplain area along a sediment-rich river might have four to six kinds of shrimp and three to four kinds of crab. Shrimp production is probably relatively high, but most of the annual increment is culled by predatory fishes, such as the croakers.

Shrimp are exploited by humans on a small scale throughout the Amazon Basin. But, compared to fish, they are of little commercial importance, except in the estuary region and in the lower reaches of the Rio Tocantins. Inland shrimping is concentrated along the Amazon River and its floodplain near Santarém. Two species are harvested. One is extremely small; the other is one of the largest freshwater species found in South America.

To what extent could the shrimping be expanded? There has been some experimentation with shrimp farming in the Amazon, and the marketability of at least two species is promising. Shrimp from the estuary, however, are considered more delicious than are upstream varieties. Estuarine shrimp are the main species, in salted form, sold in the inland cities as well. Freshwater crabs are rarely eaten in the Amazon—mainly because they are too small. Their primary use is for fish bait.

The Amazon may provide fine shrimp habitat, but it is not very favorable for shelled mollusks. Other than in sediment-rich rivers, and perhaps in the clearwater Rio Tapajós and Rio Xingu, mollusk production in Amazonian waters is severely limited by low levels of calcium and phosphorus. Despite the paucity in overall biomass, at least a dozen mollusk species inhabit floodplain waters. The most commonly seen are planorbid snails and several kinds of clams.

A good indirect indicator of snail production in particular areas is the size of the limpkin population. This large bird (which ranges as far north as Florida) has a slender, decurved bill and feeds almost exclusively on snails. Some of the planorbid snails of the Amazon attain shell diameters in excess of 15 centimeters. Though aquatic herbivores, planorbid snails climb trees or shrubs to lay eggs above flood levels in order to avoid predation by fish and other aquatic animals. Their survival thus depends on floodplain vegetation. These snails could probably be farmed. They might also serve as supplemental food for fish species raised in aquaculture.

Today's Amazonians rarely eat freshwater mollusks. Archaeologist Anna Roosevelt of the Field Museum of Natural History in Chicago has shown, however, that bivalves were exploited by Indians along the lower Amazon River, at least in the area near Santarém. Huge mounds of mollusk shells now mark ancient village sites. But the sizes of these deposits do not necessarily mean that bivalves were more important than other protein sources, and especially fish. Bivalve shells were discarded on land,

and obviously in the same general places, whereas fish bones may have been thrown in the water or scavenged by rainforest animals when left on land. From a biological perspective it is unclear exactly why, where, and how the Indians captured so many bivalves. The freshwater bivalve resource has perhaps been overlooked by modern Amazonians. It is highly unlikely, however, that natural stocks are productive enough to support large commercial operations. There may nevertheless be some potential for developing an aquaculture based on bivalves.

THE UNEXPLORED WORLD OF INSECTS

Amazonian animals are many orders of magnitude more diverse than are plants in the rainforest ecosystems. Most of these animals are insects, the majority of which have not been described scientifically. According to entomologist Terry Erwin of the Smithsonian Institution, millions of yet-unknown insect species inhabit the Amazon.

Insects are by far the most diverse group of organisms found on Amazonian floodplains. No estimate has yet been made of the total number of insect species that inhabit Amazonian floodplain forests. It is known, however, that rainforest insects as a group have a high degree of endemicity, with many species expressly adapted to vertical migrations made seasonally up and down the trees in accordance with flooding levels. As with all rainforests, most of the insect diversity surely is concentrated in the canopy.

Floating meadows are also rich in insects. Many of these species have aquatic larvae that live amid the submerged roots. The biting insects, such as some of the mosquitoes, are most abundant along the sediment-rich rivers because of better nutrient levels on which their aquatic larvae depend. Cattle ranching has probably enhanced mosquito populations, owing to nutrient-enrichment of floodplain pools through feces and an abundance of livestock as blood hosts.

Arthropod heaven.
Floating meadows provide fabulous habitat for insects, crustaceans, snails—and hence fish.

Because insects are so closely linked to human diseases, any disturbance of major biodiversity patterns must be of concern. The Amazonian floodplains have been relatively free of some of the major insect-borne diseases of the tropics, such as leishmaniasis and, until just recently, probably even malaria along many of the sediment-rich rivers. Perhaps the major health-related insect problem on floodplains is the bacterial contamination of foodstuffs by flies. Cattle and buffalo ranching has served to swell fly populations along the Amazon River. Several African fly species, which are also culprits of these infections, have been introduced and are spreading with cattle ranching. Exotic fruitflies have also been accidentally introduced into urban centers, and several species adapted to humid conditions are now spreading along the floodplains.

Overall, the great environmental changes now pressing on the Amazon River floodplain threaten to push huge numbers of insect species—the undiscovered as well as the known—over the brink of extinction. We can easily verify that howler monkeys have been eliminated from a floodplain area, but the numbers of insect species that are forced into a similar fate is beyond practical means of ascertaining. Sadly, because of limitations in funding and the paucity of trained entomologists, no more than a few hundred new insect species are being described for the Amazon each year. When discussing biodiversity, moreover, the focus is usually on plants and vertebrates. We know these two groups better than we know insects—and we seemingly want to know them better.

Birds, mammals, reptiles, amphibians, crustaceans, mollusks, and insects: what of the fish? Because fisheries offer the greatest potential of all Amazon resources for effective and sustainable development, we devote the next two chapters to this single group of animals.

CHAPTER FIVE

FISH AS OUR ECOSYSTEM EYES

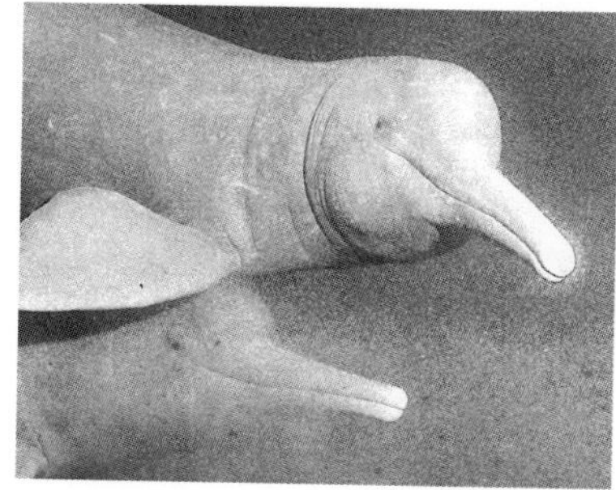

The Amazon Basin has the most diverse freshwater fish fauna in the world. Taxonomically fish are by far the poorest-known group of Amazonian vertebrates. Estimates of the diversity that might be present range from 2,500 to 3,000 species, though only about 1,700 species have been described.

An understanding of fish ecology offers opportunities to access what is probably the best general environmental signpost for charting changes in Amazonian river and floodplain biodiversity. This is not to say that other animal groups are unimportant and should not be considered. But fishes, because of their great diversity and interaction with many kinds of habitat, including rainforests, will surely yield the most information for the level of scientific effort and money expended. They can serve as our ecosystem eyes.

Unlike most other animal groups in the Amazon, fishes are of direct importance economically. The compelling need to manage and conserve valuable fisheries can be a powerful argument for preserving entire habitats on which these fishes depend—and thus the overall biodiversity associated with the floodplains. The welfare of fish has also traditionally been the main resource measurement for the effects of dams and pollution. This chapter explores the nature of Amazonian fish populations and their role as indicators of the health or degradation of the region's rivers and floodplains.

TRACKING FISH MIGRATIONS

Most studies of Amazonian fisheries have investigated markets rather than the ecology of individual species or communities over large areas. These perspectives are, however, complementary. Both are necessary in order to glean a general idea of how

Catfish migration on the Rio Madeira. *Some Amazonian catfish species migrate two or three thousand kilometers between the estuary and headwaters where they feed and spawn. The Teotônio rapids of the upper Rio Madeira is the best place in the Amazon Basin to observe these migrations. The rapids are also a favorite haunt of commercial fishermen.*

the fisheries function ecologically. The seasonal migration of a large number of species of commercial importance, for example, can be used as a gauge for developing good management and conservation programs.

Almost all Amazonian fishes change habitats to some extent during the course of the year in response to fluctuating river levels. In making these seasonal shifts, many species undertake long-distance migrations. At the local level these migrations are well known to commercial fishermen and to scientists who have studied them. Indeed, the most important fisheries in the Amazon take place during the times and at the locations when the target fishes are congregated or most vulnerable.

Although the general scope of these migrations is well known, neither fishermen nor biologists have a clue as to the geographical limits of discrete stocks—that is, genetically isolated populations. This type of information would be useful in determining the degree of overlap in the stocks harvested in different fishing regions. In a system as large and open as that of the Amazon Basin, and with most of the commercial fish species widely distributed, there is no easy way to determine how many distinct populations, or stocks, of a given species might be present throughout the range that is fished.

Taxonomic knowledge in some cases is still too poor to determine even whether the type of fish taken in one region is truly the same species as that taken in another. Can the two populations interbreed? If one stock is driven to extinction, can the other replenish the loss, or is a distinct species lost forever?

None of the stocks of the most important commercial species are restricted to individual river basins. Fishermen, in fact, often follow schools from one tributary to another during seasonal migrations. Possibly some of the migratory species found in the Amazon lowlands are not divided into several or many stocks, but exist rather as one large, interbreeding population. Genetic studies will, however, be needed to test this hypothesis.

The principal migratory food fishes in the Amazon Basin are the characins and catfishes. Other groups, such as herrings and drums, also make seasonal shifts, but the movements of these fishes in the river channels are more difficult to observe. Migratory patterns in the central Amazon lowlands are quite different from those that have been observed in headwater areas. In most of South America, if not in the world, fishes tend to migrate upstream once a year to spawn in the upper courses of the main river or its tributaries. This also happens in Amazonian headwater areas. Perhaps some of the large catfishes that annually spawn in the headwaters begin their journey from as far downstream as the tidal estuary. Most of the other species in the Amazon lowlands, however, migrate upstream or down at least twice annually.

The most important migratory fishes exploited in the commercial fisheries of the central Amazon spawn only in the sediment-rich rivers—though these species, at least as adults, may live mostly in blackwater and clearwater tributaries. Spawning takes place when water level is rising rapidly and the floodplains are being inundated. Exact spawning sites are usually not known because muddy water makes it extremely difficult to observe these fishes. Many studies, however, have verified that the floodplains of the sediment-rich rivers are the principal nursery habitats of the migratory characins and probably of the smaller catfishes as well.

Floodplains of sediment-rich rivers provide the most complex fish habitats in the Amazon Basin. Owing to variations in local drainage, both blackwater and clearwater lakes may be present on the floodplains of sediment-rich rivers. The floodplain of the Amazon River itself is a good example of this. Areas inundated by the nutrient-rich waters of the Amazon River usually support an exuberance of floating meadows that, under natural conditions, fringe huge areas of flooded forests. When sediments settle

An extraordinary migrator. *The dourada catfish, which migrates from the estuary into headwaters, demonstrates the very large areas that must be considered when designing management programs for commercial fisheries. Overfishing in either habitat can damage the fishery in the other.*

in the calm waters of the floodplains, transparency improves and plankton production is heavy in open waters exposed to sunlight. In turn, floating meadows and the open-water plankton communities of nutrient-rich floodplain lakes are important for young fish. This fact largely explains why migratory species move down the clearwater and blackwater tributaries to spawn near habitats inundated by sediment-rich rivers in flood.

Unlike anywhere else in the world, many Amazonian fishes depend on fruits and seeds that fall into the waters from the natural orchards of the flooded forests. As adults, many of the migratory species thus have special adaptations for feeding on the fruits of the floods. The tambaqui, for example, has huge, crushing molars—making its dentition look more like that of a horse than a typical fish.

The advantages gained by fishes that can use several or many different kinds of habitat during their life histories is perhaps the principal factor that explains why the Amazon is rich in migratory species. Such adaptations, however, mean that migratory fishes depend on enormous areas of the main river and its tributaries—a fact that can compound management problems.

The migratory characins of the central Amazon are unusual in that they make two migrations each year, or at least most do in most years. The second annual migrations take place sometime between the time the floods begin to ebb and the end of the low-water period. Schools of characins move en masse out of the floodplain areas of the sediment-rich rivers, into the main channels, and thence upstream. Other schools are easily detected moving down the blackwater and clearwater tributaries and then up the sediment-rich rivers such as the Amazon, Rio Madeira, and Rio Purus. Most species eventually enter another tributary system or floodplain lake area of the main river, further upstream, though the exact distances of these upstream movements are not known. In some cases, schools may enter a tributary for a few days or weeks and then move back downstream to the sediment-rich river where they again migrate upstream.

At their greatest intensity during the low-water period, these second annual migrations in the central Amazon are generally known as the piracema, which is an indigenous word meaning "fish-above," in reference to the main direction of movement. During the lowest water period these schools are very easy to observe because of jumping activity and dolphins chasing them. It is this low-water concentration of fish that has reinforced folklore beliefs in superabundance. This folklore belief, in turn, bodes ill for any hope that local peoples, without the intervention of government agencies, could self-regulate their takes in their (and the fish's) own best interest.

The ecological purpose of the second annual migrations appears to be an adaptation to counterbalance the downstream displacement of newborn fish. Most of the migratory characins appear to spawn in or along the main river channel, and the evidence to date suggests that their larvae are carried downstream by strong currents. There may also be a behavioral component involved in this downstream displacement, but young fish have yet to be observed well enough to determine whether they are swimming or merely drifting.

In effect the second annual migrations place characin fish schools in double jeopardy because fishermen can exploit them heavily twice a year. Government authorities in the Amazon have tended to believe that the exploitation of spawning schools puts characin stocks in most danger of overfishing. Market catches in the Amazon, however, seem to indicate that far more tonnage is captured during the low-water season, when schools are more concentrated and easier to find in the river channels.

In addition to the commercially important characins, large catfishes constitute another valuable fisheries. Large catfishes are widely distributed in the Amazon Basin. They use the tidal estuary as nursery habitat. The estuary is a highly productive region because of plankton production, detritus accumulation, and extensive tidal forests. Adult and preadult schools of these estuarine species migrate upstream. Commercial interests exploit them all along the way—from the Amazon River mouth to as far upstream as Peru and Bolivia.

The migratory catfishes fall into two ecological groups. Those of the family Pimelodidae are predators. These large catfishes are the most abundant predators of fish in river channels. By migrating upstream they are able to take advantage of prey that become concentrated during the low-water period or during spawning runs. Ecologists refer to these movements as trophic migrations. Ronaldo Barthem of the Goeldi Museum in Belém has discovered that some of the large catfishes spawn in the western reaches of the Amazon Basin. Their newborn then drift or migrate downstream to the estuary, which they use as nursery habitats.

The second group of migratory catfishes comprises at least two large, armored species that are primarily herbivores. Unlike the pimelodid catfishes, the armored catfishes consume the fleshy fruits, leaves, and mollusks of the tidal forests. Adults migrate up the Amazon River and into many of its tributaries, apparently to reduce competition with their young, who remain in the estuary.

ENVIRONMENTAL CONSEQUENCES OF THE FIVE DAMS

The program of Amazonian dam construction should have included broad-scale investigations of fish migrations before any of the impoundments were closed. In no case did any of the researchers working in hydroelectric projects do surveys from headwaters to mouth in order to develop hypothetical models of fish migrations and movements within and out of the individual rivers involved. Had these studies been conducted, environmentalists and the public outcry might have forced officials to consider the need for fish ladders or other expensive devices—which would have exacerbated already inflated budgets.

The five dams thus far constructed are in the coastal, central Amazon, and Rio Tocantins areas. The first, the Coaracy Nunes dam on the Rio Araguari in the coastal state of Amapá, was finished in 1975—before Amazonian projects had attracted much attention from environmentalists. No ecological studies of fish were conducted. Officials did not even undertake a general taxonomic survey of the species present before and after the dam was closed.

Migratory schools apparently still move up the Rio Araguari at the beginning of the rainy season, and these may be spawning below the dam. Owing to the Rio Araguari's proximity to the ocean, the tidal effect is felt as far as the dam. The tidal influence on fish migrations thus may be important in this system. Migrations upstream of the dam have been observed in headwater streams, such as the Rio Cupixi. There are still enough old fishermen in this area that an extensive series of interviews, from the headwaters to the mouth, could probably piece together the general fish migration patterns before and after the Coaracy Nunes dam was closed. Another way to assess the fisheries consequences would be to compare fish populations and movements on the Rio Araguari with those in the nearby, free-flowing Rio Amapá.

The main factor to be considered in the Amazon in connection with dams is not whether there are migratory fishes—beyond any doubt there are—but the size of each river dammed and its importance to the local fisheries. Specifically, what proportion of the characin food fishes are now prevented twice annually from getting downstream or upstream, to spawn and disperse, in the blocked tributaries?

A study of fish migrations in the lower reaches of the Rio Tocantins was conducted in 1981 and 1982—that is, before the Tucuruí dam was closed in 1984. The study found that most of the commercial catch of the lower Rio Tocantins in the pre-Tucuruí period consisted of two species, the plankton-feeding mapará catfish and a detritus-feeding characin called a curimatá, which resembles a carp. Immature and adults of both species were found to migrate upstream.

The mapará catfish apparently did not migrate much farther upstream than Tucuruí, that is, to about the first rapids. Juvenile mapará were concentrated in the lower Rio Tocantins for most of the year. During the lowest water period, however, when brackish water may occasionally enter the lower Rio Tocantins, juveniles moved upstream as far as the rapids. Adults migrated up to about this point, too, beginning in November in order to spawn at the onset of the rainy season when river level was rising rapidly. The newborn then either migrated back downstream or were carried by the current to the lower part of the river, which is their main nursery. Adults also dispersed downstream after spawning.

What about the curimatá? In South America in general curimatá, of which there are many species, are the most important migratory food fishes. Their long-distance migrations have been studied in several different river systems in southern Brazil, Argentina, and Bolivia. Migration patterns derived from fisheries observations were also outlined for the Rio Madeira, a sediment-rich Amazon River tributary to the west of the Rio Tocantins. Curimatá migrations in the Rio Tocantins are more similar to those of southern South America than to patterns that have been observed in the central Amazon. In the Rio Tocantins curimatá migrate upstream to spawn. In contrast, curimatá that live in clearwater tributaries of the central Amazon migrate downstream to spawn in the turbid rivers, such as the Rio Madeira and Rio Amazonas. This observation of habitat-specific differences in fish behavior points out how important it is to consider individual river basins in the Amazon in order to understand fish migrations.

Unfortunately, the limited study of curimatá on the Rio Tocantins did not address whether populations residing in the middle and upper stretches of the river ever used the lower part (that would be closed off by the dam) as nursery habitat. Curimatá are famous for their ability to pass rapids, and thus their migrations could have been observed easily from shore. But the observation was not made, and it is too late now.

Once the Tucuruí dam was closed, and reservoir conditions became unfavorable because of low oxygen levels and perhaps acidity from decaying vegetation, curimatá and other migratory species were apparently forced upstream. This upstream concentration of fishes likely explains why commercial catches increased in the middle and upper Rio Tocantins after the dam was closed.

Jaraqui, another fish resembling but not related to carp, is closely related to curimatá. It was probably the most important commercial migratory species in the Rio Curuá-una, Rio Jamari, and Rio Uatumã. In the years after Curuá-una was impounded, jaraqui schools were observed just below the turbines in what was apparently an attempt to migrate upstream. Because jaraqui often stop moving upstream when reaching cataracts, and because the dam was built at a former cataract, we cannot know to what extent the dam actually interferes with natural movements.

Some of the most abundant species captured in the Curuá-una reservoir after the dam was closed are known to be schooling fishes and to migrate. At least two of these species either migrate upstream to spawn or they have adapted to reproduce in the reservoir. We know this because young size classes were captured in the reservoir during fishing experiments. Whatever the effect of hydroelectric projects on these and other stocks of fishes, we can surmise that each of the five dams effectively cleaves one larger population into two. Only time will tell what evolutionary effects may ensue.

Although the first of the five major dams was constructed without an ecologic or taxonomic study of the affected fish, construction of each of the later four dams was accompanied by limited fish inventories. The National Institute of Amazonian Research in Manaus was contracted during the late 1970s and 1980s by Electronorte, a branch of Brazil's state-owned energy monopoly, Electrobrás, to do faunal surveys of the Rio Curuá-una, Rio Tocantins, Rio Uatumã, and Rio Jamari. In no case were sufficient funds or personnel available to adequately survey an entire watershed. Most of the studies were restricted to areas relatively close to the dam sites. The most intense survey was done on the lower Rio Tocantins. There, approximately 325 fish species were collected. In our estimation, those 325 species do not even begin to represent the actual faunal assemblage of the entire area affected by the Tucuruí dam.

A comparison of experimental gillnet catches before and after Tucuruí was closed revealed a 49 percent reduction in fish diversity below the dam and a 50 percent reduction above. The downstream decrease in diversity was at least in part attributable to the disruption of the migratory cycle of many species. Diversity in the area filled by the reservoir fell 55 percent once the lake filled. Species that were previously associated with rapids were (not surprisingly) absent in experimental fishing efforts once the dam was closed.

Total fish biomass as indicated by these experimental gillnet catches fell dramatically downriver of the dam. This loss was also marked by a 65 percent reduction in commercial fish and shrimp catches. These downriver fisheries are apparently still on the decline.

In contrast, there was a dramatic increase in fish biomass in the waters upstream of the dam. The fisheries yield grew by 315 percent between 1984, when the river was still free-flowing, and 1989, by which time a large reservoir was formed. Total annual catches jumped from about 325 to 1,425 tons. The two most important fish groups harvested in Tucuruí reservoir during this period were the tucunaré cichlids and the croakers. Both are predators that probably benefited from the relatively transparent waters in the reservoir lake. Tucunaré and croakers hunt visually, and thus the greater transparency in the dam-lake waters might give them an advantage. Increased transparency also led to greater algae production and probably an increase in the small prey species on which these predators feed. Then, too, Tucuruí reservoir still contains huge quantities of organic matter from the drowned rainforest that is now in various stages of decomposition. As this material decomposes, nutrients are released into the water, leading to greater algae production. This nutrient recycling will continue for another decade or two until the drowned vegetation is completely decayed. Once this happens, productivity will probably begin to fall, since the Rio Tocantins is a relatively nutrient-poor river.

The fish fauna in the vicinity of the Curuá-una dam has been surveyed three times: in 1977, when the dam was closed in 1978, and again in 1982. A total of 214 fish species was captured in the aggregate of these three counts, but the data thus far collected are insufficient to determine if there has been any decline in diversity. The data available do, however, suggest that there was a replacement in the most abundant species, at least as revealed by gillnet surveys. Algae-feeders, predatory piranhas, and a pelagic catfish seemed to benefit from the reservoir, whereas some of the medium-size characins became rarer.

Macrophytes, or floating herbaceous plants, thrived in the Curuá-una reservoir after the dam was closed. By 1979 aquatic plants covered more than a fourth of the reservoir. Floating plants in dam reservoirs are a major concern because they act as pumps (transpiration) that lead to serious water loss. In the first couple of years after impoundment, nutrient levels in the reservoir behind the dam increased because little water was released as the artificial lake filled. The high nutrient level largely explains why the floating plant population exploded. When flow-through was expanded in 1979 to generate more power, the nutrient levels that had accumulated in the reservoir were diluted, and macrophyte populations began to decline.

All Amazon dams thus far have been built at the first major rapids of a river. Ichthyologists have long been aware that Amazonian cataracts have unique species and that the faunas in most tributary systems begin to change above the first rapids. Furthermore, there is much species endemicity in individual river systems. Studies of cataract faunas were not undertaken within any of the rivers dammed. Nor were comparisons made with other tributary systems to determine if the same species were involved. In nearly all major fish collections made near cataracts or in rocky areas, species unknown to science were found. It is thus highly probable, though almost impossible now to prove, that Amazonian dams have indeed resulted in species extinctions, and especially of highly adapted species that lived in the rock crevices of now-flooded cataracts.

FLOODPLAINS, FOOD CHAINS, AND FISH PRODUCTIVITY

Floodplains contain the most productive habitats of Amazonian inland waters. A combination of flooded forests, floating meadows, and algae communities in open waters form the base of the main food chain sustaining fish. The relative importance of each of these three habitat components of the food chain depends largely on water type and the amount of deforestation that has taken place.

Floodplains of the inland waters of the Amazon have a much higher fish diversity than either their adjacent river channels or the tidal estuary. Not surprisingly the greatest diversity in food fishes is also taken in floodplain waters. More than fifty species of food fishes are regularly captured in most heavily exploited floodplain waters, and perhaps up to a hundred species, if occasional catches are included.

Other than the migratory species that form large schools and are easily observed, it is not known to what extent the other food fishes are resident in any particular area. Even nonmigratory fishes probably wander considerable distances during the floods in search of food, breeding habitats, or better oxygen conditions. Sound policies for fisheries management depend upon knowledge of the extent to which fish populations remain isolated in individual floodplain lake areas. This is especially true if management area boundaries coincide with individual floodplain lakes. These lakes are easily defined during the low-water period when they become isolated by land from their upstream and downstream counterparts. Nevertheless, the extent to which these lakes might have discrete fish stocks is still unknown.

Floodplain fish communities are not fixed mixtures. Instead, the kinds and proportions of fishes are ever changing because of fluctuating water levels and migrations. During the floods most species move into the inundated forests to feed, breed, or seek greater protection from predators. Other species prefer the floating meadows until these plant communities die back during the low-water period. A lesser number of species is specialized to feed in open waters. There the fisheries comprises plankton feeders, detritivores, or predators, and all three types are represented among the food fishes. Overall, floodplains—not the river channels—are the most important nursery habitats for the majority of food fishes of inland waters. With the depletion of adult stocks in the last decade, the younger size classes are now being heavily exploited in floodplain waters.

The Amazonian aquatic environment and its biodiversity are far too complex, and research funds far too scarce, to design realistic quantitative experiments to calculate total fish productivity. We do know, however, that the three main sources of food energy that sustain Amazonian fish communities are floodplain forests, floodplain herbaceous plants, and open-water algae communities. The relative importance of each of these sources varies greatly, depending on river type and the degree of floodplain deforestation. The productivity of planktonic algae is not difficult to measure. For this reason, more than its implied importance, algae productivity has been relatively well studied in central Amazon floodplain waters. Fish productivity, however, cannot be calculated from plankton production alone. Furthermore, many Amazonian fish species have life histories that tap energy derived from two or all of the three major sources of primary productivity. This is especially true of some of the detritus-feeding species. The detritus they eat could be derived from woody plant parts, floating meadows, or algae. A further complication is that many fishes undergo seasonal migrations between different habitats.

In the heavily deforested floodplain area of the Amazon River between the Rio

Negro and Rio Purus, Peter Bayley of the Illinois Natural History Survey estimated fish production to be 1 percent of primary production. Primary production is a measure of the total carbon fixed in plants through photosynthesis; fish production, in turn, is the proportion of that initial allotment of fixed carbon that is reshaped into the carbon of fish tissues. Using the data available, Bayley calculated the following approximate contributions from the three main sources of primary production: algae (7%), herbaceous plants (69%), and flooded forest litter (24%). Since the area Bayley sampled is heavily deforested and only litter was considered, the estimated carbon contribution from floodplain forests would be considerably higher under natural circumstances.

In an experimental study, Carlos Lima and Bruce Forsberg of the National Institute of Amazonian Research attempted to determine the source of the carbon incorporated into some fish species. They discovered that algae had to be of much more importance than indicated in Bayley's estimates. In any case, Bayley's estimates suggested that the local fisheries were taking less than 3 percent of total fish production in the region studied, or somewhat more than 7 percent of the species that can grow to at least 25 centimeters in length.

The second most voracious predator of fish in the Amazon. *The bôto dolphin, unlike the smaller tucuxi dolphin, readily leaves the channels and lakes during the high-water season to pursue prey into the flooded forests. Nearly blind, the bôto dolphin is well served by a sophisticated sonar system that penetrates muddy waters. Unlike the manatee, it has not been hunted for food, and today can often be seen surfacing in river channels, lakes, and flooded forests. As a predator of fish, it is surpassed only by people.*

Even if Bayley's harvest estimates are far too low, they still suggest that there is a large fish biomass to be exploited if it could be captured economically and marketed. Since the Amazon Basin is not a homogeneous system, Bayley's pioneering fish production work needs to be replicated in other areas. Comparisons might then be made in order to assess how much more fish could be harvested without destroying the fisheries. There is a danger, however. Could harvests, in fact, be increased without endangering young size classes? This side effect of greater takes does indeed seem to be a real threat.

Fishermen do, of course, compete with other predators, of which there are a large number of species in the Amazon. Caimans are reptilian examples, but the two large caiman species of the floodplains are now very uncommon. Piscivorous birds can be important predators, too, though their populations have dwindled in the Amazon Basin because of egg collecting and hunting.

The only important mammalian predators of fish in the floodplains, river channels, and lakes are the dolphins. Dolphins are utterly dependent on fish as prey. Amazonian dolphins rely entirely on their sonar for finding and pursuing prey, as they are nearly blind. They are the most intelligent and (because of their high metabolisms) the most voracious predators in Amazonian waters. A fifty-kilogram dolphin probably eats twenty or even thirty times more prey than does a predatory catfish of the same size. The largest of the two species is the bôto dolphin, whose snout is so long that it superficially resembles more the prehistoric ichthyosaurs than its own mammalian kin. It can reach 2.5 meters in length and weigh 150 kilograms. The bôto dolphin, unlike the smaller tucuxi dolphin, readily leaves the channels and lakes during the high-water season to pursue prey into the flooded forests.

The main predators of fish in the Amazon are other fish. A study done in the floodplain of the lower Rio Solimões near Manaus indicated that at least 75 percent of the production of juvenile and adult fish up to 24 centimeters long and of shrimp was consumed by piscivorous fishes. Piscivorous fishes in central Amazon markets account for only about 5 to 20 percent of total catches. The situation is quite different in the tidal estuary, where predatory fishes make up nearly the entire commercial catch and the largest percentage of the local, subsistence landings.

From a food chain perspective, central Amazon fisheries are in one way unique. Unlike anywhere else in the world, fruit- and seed-eating species constitute a major share of the annual catch. In the mid 1970s, for example, adult tambaqui, which are mostly fruit and seed eaters, accounted for more than 40 percent of the total catch landed in the main market in Manaus. Even in the upper Rio Madeira region, where floodplains are much smaller than those of the Rio Solimões and its tributaries, fruit-eating fishes represented about 40 to 50 percent of the commercial catch sold in Porto Velho (Rondônia) between 1984 and 1989. Among these, the tambaqui has been and is still the single most important commercial species in the central Amazon. Now even young size classes of the tambaqui (adults of which may reach a meter in length and thirty kilograms in weight), which feed heavily on fruits as well as zooplankton, are beginning to be heavily exploited.

After the fruit and seed eaters, detritivores are the next most commercially important kinds of fishes. Detritus is a catchall term for organic material in some stage of decomposition. Detritus feeders necessarily also consume inorganic matter and small living organisms found on or mixed with the decomposing material. Detritus-feeding species account for about a third of the commercial fish yield of the central Amazon. Nearly all important detritivorous species are migratory characins (though a large doradid catfish is sometimes locally important, especially in the Santarém area of the Rio Amazonas). Amazonian detritivores have traditionally been considered second-class food fishes because of their muddy taste and the large number of spines that the characins have. Nevertheless, many species could undoubtedly be more intensively, yet sustainably, exploited.

Hundreds of Amazonian fish species feed on detritus either exclusively or seasonally when preferred foods are unavailable. The exact carbon sources of this detritus are not yet known. But experimental evidence suggests that the characin detritivores cannot sustain themselves by digesting only the small living organisms found in the detritus. They must also be able to digest the decomposing material.

One experiment indicated that the organic carbon upon which Amazonian detritivorous fishes depend originates with algae. Many, if not most, of the detritivorous food fishes of the Amazon fatten in the seasonally inundated floodplain forests. If decomposing algae is ultimately the main carbon source for these detritivores, then flooded forests must play an important ecological role in accumulating and concentrating this organic material. Whether flooded forests themselves supply sufficient nutrients to foster rich algal growth is unknown. Instead, perhaps these inundated habitats, because of their geometrical complexity, act as a giant "food net" that traps algae and detritus carried into the flooded forests from open waters.

Although the few most important commercial food fishes of the Amazon are easy to classify ecologically, the majority of species sold in local markets are not. They cannot be classified simply as frugivores, detritivores, predators. Rather, they are om-

Favored fish habitat. *Amazonian fishes evolved with flooded forests and floating meadows. They depend on these habitats for feeding and breeding. Some of the most important commercial species feed only in the flooded forests, fasting in river channels and lakes during times of low water. Flooded forests also provide seasonal protection from fishermen.*

Fish food of the flooded forest. *Fruits, seeds, insects, and spiders that fall out of the trees are important food groups that support many species of fish in floodplain forests. The seeds of fleshy fruits swallowed by fish often pass through undigested. Seed dispersal is thus a mutualistic relationship between the floodplain flora and the fish species that favor them.*

nivores, feeding on a wide variety of foods. These omnivorous species will become more important in the fisheries economy as the preferred taxa become rare.

Only one scientific estimate of the total annual yield of Amazonian fisheries is available. Peter Bayley and Miguel Petrere estimated that two hundred thousand tons are landed annually in the Amazon Basin. This estimate was based mostly on data from the late 1970s and early 1980s—the only period when several markets were monitored simultaneously and independently. Bayley and Petrere designated landings in most urban centers other than Manaus as subsistence, rather than commercial, fisheries. Based on that perspective, about 60 percent of the total yield was accounted for by the smaller urban centers and subsistence peasants scattered along the rivers. The remaining 40 percent was taken by the Manaus and Belém fleets. If data were available that distinguished the many towns and cities with populations greater than five thousand, so that they did not have to be grouped with the subsistence population for statistical convenience, then the total yield from urban centers would probably exceed 75 percent, rather than the 40 percent the first estimates revealed.

GOLD MINING AND THE DANGERS OF MERCURY-CONTAMINATED FISH

Scientific and political concern regarding the general dangers of mercury contamination arose only after major ecological disasters and human deaths, such as those at Minamata, Japan, in the 1960s and in Iraq in the 1970s. Serious environmental and human contamination by mercury pollution has also occurred in Pakistan, Sweden,

Guatemala, Colombia, Eastern Europe, and now Brazil (states of Bahia and Alagoas). In most of these cases, contamination was caused by the discharge of industrial or hospital wastes, through which mercury entered food chains. No one has yet satisfactorily compared the Amazon environmental situation with previous cases of mercury pollution and contamination.

Almost all stocks of mercury used in Brazil are imported. Until 1984 Mexico accounted for more than 90 percent of that supply. Since 1985, however, nonproducers (that is, traders) of primary metallic mercury, and especially the Netherlands, Germany, and the United Kingdom, became the major suppliers to Brazil. Mercury importation into Brazil peaked in 1989. This peak reflected the peak of the Amazon gold rush. Commercial imports of mercury for gold mining have now surpassed industrial demands.

Amazonian wildcat miners are now the principal consumers of mercury in Brazil. These miners depend on a process called amalgamation to concentrate and remove gold from deposits washed in pans, sluices, and other devices. Metallic mercury is the additive used to chemically amalgamate, and thus capture, the tiny grains of gold. The gold-and-mercury amalgam is then isolated, and the two metals separated, by applying a direct flame from a blowtorch. Since mercury has a lower temperature of volatilization than other metals, it evaporates and leaves behind the gold and other ores to which it amalgamated.

Mercury is released directly into streams and rivers when not all the mercury slurried with the ore is taken up by another metal, such as gold. The volatilization phase is the other port of entry for mercury into the environment. Research has only just begun on mercury pollution in the atmosphere over the Amazon. Atmospheric mercury pollution could be a serious problem, since gold mining activities release huge amounts of it in vaporous form. Because the Amazon is a rainforest, that mercury soon precipitates and atmospheric pollution becomes water pollution.

Mercury, at about $40/kg in the Amazon, is just a small part of gold-mining investments. For comparison, mercury on a per-weight basis costs less than 1 percent of the price at which gold is sold. Because mercury is relatively cheap, miners have little incentive to recover and recycle it. Various agencies have developed, or promoted the use of, retorts to recapture mercury during the burning off of the amalgam and the concentration of ore. These retorts are already being sold in the Amazon, but the low price of mercury, and the inability of the government to control its use, suggests that the alternative technology will not be widely adopted, at least in the short run.

The amount of mercury used in the amalgamation process in Amazonian gold mining appears to vary from site to site, and from operation to operation. The rate of sediment flow through sluice boxes and the type and quantity of gold seem to be the most important factors determining the amount of mercury used. Most calculations suggest at least a 1:1 ratio of mercury to gold extracted. The upper Rio Madeira region in the state of Rondônia is reported to have the greatest mercury use, at 1.3 parts to one part of gold extracted. Thus, as much as a hundred tons of mercury may have been discharged into Rio Madeira waters between 1974 and 1985.

Once mercury enters the environment, and thence food chains, it is not rapidly withdrawn by natural processes. The toxic effects linger, cycle, and spread. Globally, the annual increment of new mercury pollution varies from 630 to 2,000 tons. Brazil is thus responsible for perhaps 2 to 11 percent of all mercury discharged into the global environment each year. And it seems that Amazonian wildcat miners are now responsible for most of Brazil's mercury pollution.

Both the direct input of mercury to waters and its precipitation from volatiles can lead to contamination of river and lake sediments. Sediments thus provide the prime chemical pathway for the introduction of mercury into food chains. Only metallic mercury is used in gold mining. This type of mercury, however, can enter food chains only after it has been converted into a quasi-organic form, methyl mercury. This usually happens when the mercury is absorbed by microorganisms—a process called methylation.

Methylation can occur in both oxygenated and anaerobic situations. Anaerobic methylation might, for example, take place in the organic sediment layer on the bottoms of rivers. Once mercury is methylated, however, it is able to traverse cellular membranes of the animals that consume the detritus or flesh that contains it. Methylmercury is also soluble in certain kinds of body fats, thus facilitating its concentration in animals. Owing to the ecological processes of bio-accumulation or biomagnification, mercury will become more concentrated at each higher step in the food chain. At some threshold it reaches levels dangerous not only to life in the rivers and forests but to humans who consume fish and other foods harvested in contaminated environments. The collapse of the Great Lakes fishery in the United States for health reasons—despite an abundance of fish—is a sobering example of what could perhaps happen in the Amazon.

High concentrations of organic acids make it even easier for mercury to enter food chains. Because of the large number of blackwater rivers carrying organic acids, mercury pollution should be a particular concern in the Amazon Basin. Many clearwater rivers also have high levels of organic compounds. Furthermore, these rivers have low sediment loads and thus less capacity to absorb or dilute mercury contaminants. In 1992 much of the middle and upper Rio Negro—a blackwater river—was invaded by wildcat dredge operations. Because of its high acidity and minimal suspended load, the Rio Negro may become the mercury-contaminated disaster that many in the 1980s feared the muddy Rio Madeira—which has been the principal focus of studies of fish contaminated with mercury—would be.

Keeping fish populations free from mercury contamination should be a top governmental and cultural concern in the Amazon because of the importance of the local fisheries in supplying cheap animal protein. And fish has been the main pathway leading to human contamination in other parts of the world that have suffered from mercury pollution. Although mercury pollution in the Amazon has been recognized as a potentially serious problem for at least a decade, little is known about its concentration in food chains.

Mercury is known to bond to suspended inorganic particles and thus can be moved vast distances by currents, shifting pollution problems to places far downstream of the locus of entry. A neutral to basic pH seems to decrease the uptake of mercury in aquatic food chains. Both the Rio Madeira and Amazon River have a nearly neutral pH, and both carry a lot of suspended particles. The Rio Madeira, in particular, is heavily laden with sediments, as its headwaters drain the highly erosive Andean slopes. These two factors, along with the huge volume of water it discharges, have probably "saved" the Rio Madeira region from becoming one of the great environmental tragedies of the twentieth century. But there is still cause for worry. Too few studies have been done to assess whether disaster will yet come to the Rio Madeira if current mining practices continue. For example, recent evidence presented by Ana Boischio and Antonio Barbosa indicate that riparian peasants near Porto Velho along the upper Madeira do indeed have alarming levels of mercury in their bodies, and this is directly correlated with contaminated fish on which they largely depend.

Mercury pollution and contamination can be detected and assessed in several ways. The most common methods involve measuring levels in the water, atmosphere, fish, and in human hair, blood, and urine. Measurement of mercury levels in Amazonian waters has thus far not revealed any distinct patterns that can be linked to more general contamination possibilities. Olaf Malm, a leading mercury expert, suggested that 55 to 60 percent of all mercury discharged in the Amazon Basin enters in vapor form. It is very expensive to measure atmospheric mercury levels, and only now are adequate laboratories under development in the Amazon region. Available data are mostly from sites near gold-processing shops in towns and cities, where direct contamination takes place when mercury fumes are inhaled. Gold-processing shops tend to be concentrated along one or two streets in each town or city.

Once methylmercury is ingested in contaminated food it is absorbed by the gastrointestinal tract and is then assimilated throughout the body. In humans it can pass to the fetus. Further contamination can occur later by way of mother's milk. As much as 10 percent of what is ingested can lodge in the brain, inducing irreversible neurological damage.

Symptoms of methylmercury poisoning are similar to those of the early stages of malaria—that is, nausea, fever, and chills. The almost epidemic scale of malarial infections in wildcat mining camps and towns thus frustrates detection of mercury poisoning. Fully developed symptoms of mercury contamination in humans include motor and speech difficulties, constriction of the visual field, altered sensitivity to stimuli, loss of hearing and sight, spasms, paralysis, and even coma. Children born to mercury-contaminated mothers suffer paralysis, mental retardation, and other abnormalities.

Overall, changes in the sizes of fish populations and the health of individual fishes within them can alert us to environmental deterioration in the Amazon. Food fishes harvested for subsistence or sale can be sampled for mercury contamination. Migratory fishes that concentrate during spawning can easily be censused, thus providing a general measurement of the health of those populations and that of the vast habitat through which they may range at other times of the year. But Amazonian fishes are not just our ecosystem eyes; as food they provide sustenance for a great part of the regional population—the subject of the next chapter.

CHAPTER SIX

FRUITFUL & FRIGHTENED FISHERIES

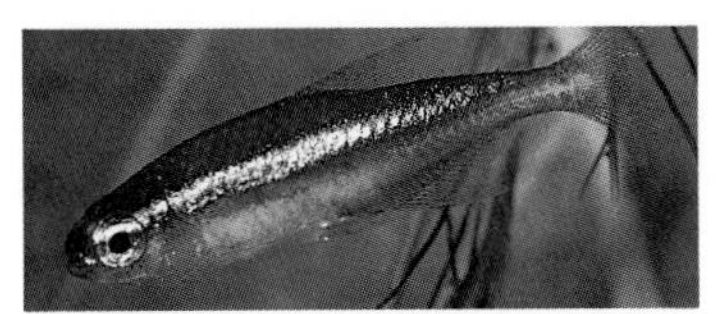

The forest floor and canopy of the Amazon rainforest is surprisingly depauperate in harvestable plants and animals. The very richness of the rainforest means that evolution has had a long time to equip plants with chemical weapons to deter herbivores. Likewise, the rainforest is not a natural farmyard because most of the animal protein is unacceptable as food. Insects, after all, are not usually looked upon by humans as a food, and nowhere in the Amazon are they a primary source of protein.

It is thus to the waters that we must turn. Of the food animals accepted by humans, fishes are by far the most successful and productive of the Amazon rainforest. Fisheries did, in fact, provide the single most important source of animal protein exploited by humans during the colonial settlement of the Amazon Basin. It has remained so ever since.

Earlier, the great variety of fishing methods developed by the Indians in part reflected the wide variety of habitats present and the richness of the fish fauna. Although scientists have gained a fairly thorough understanding of how Indians classified and exploited plants, studies of fish resource use by native groups have proved difficult. This difficulty is mainly due to species identification problems and the limited background information available on the ecology of the aquatic resources. Most Indian groups are now found in headwater regions far away from the most productive areas in the Amazon Basin, and it is in these areas where we might still gain some understanding of their interaction with fish resources. Anthropologists, such as Janet Chernela, have been able to give us a glimpse of the key role played by fish in these societies.

With the decimation of the Indians and their miscegenation with European and African peoples, peasant societies evolved in villages, towns, and cities along Amazonian rivers. These *caboclo* societies, as they are called in Brazil, effectively replaced the human ecological niches of the Indian tribes. Much of peasant knowledge about fishes and fishing was inherited, in many circuitous ways, from indigenous technology. The single most important Indian technology adopted was the dugout canoe, which provided the maneuverability and stealth required in subsistence fishing. The bow and arrow became standard equipment in peasant fishing, and it continues to be used even in commercial fisheries. Europeans introduced the first metal hooks into the Amazon Basin, but many of the baits, and especially the wide variety of fruits and seeds used, were based on indigenous knowledge about fish-feeding behavior.

Along with the technological devices adopted from the Indians came some of the names that the Portuguese and Spanish would use to refer to the fishes in their new home. Iberians were little prepared zoologically or linguistically to confront a fish fauna as unfamiliar and diverse as that of the Amazon Basin. Where they tried, such as calling some of the freshwater species sardines, they left a nomenclatural legacy that still causes confusion to the layperson. There are no standardized Portuguese or Spanish names for the fishes in the Amazon Basin. It will be a daunting task to standardize names because of the cultural and biological diversity in the region. And, if they are to be standardized, as they should be, it will have to be done rather arbitrarily because of the regional differences that exist.

THE MYTH OF SUPERABUNDANCE

José Veríssimo, the great Brazilian literary naturalist and historian, went so far as to state that the principal reason the Portuguese were able so easily to settle the Amazon Basin was that fish were available almost everywhere. Seasonally and regionally other animals, and especially turtles, manatees, and terrestrial game, were also important, but there is no evidence that any of these ever contended with fish as the main animal protein source in most of the Amazon lowlands. Beyond any doubt, fish constitute a much larger biomass than do any other food animals in or along rivers. The great diversity of the Amazon fish fauna and the large number of species at various levels of the food chain are also key factors explaining the dominance of fish as a food resource. Amazonian fishes have also been successful in tapping into the productivity of forest canopy—as they have evolved adaptations for finding and digesting fruits, seeds, invertebrates, and detritus that fall into the waters of the seasonally flooded forests. This evolutionary trend has opened an enormous habitat to them.

In the postconquest period the Amazon was settled by extractivists who, over the centuries, developed and promulgated the belief and cliché of superabundance. Entrepreneurs operated as barons, and they controlled what were essentially river and floodplain fiefdoms, with peasants living there as debt servants. These economic relationships still exist today in much of the rural Amazon. Owing in large part to the rich fisheries, peasants did not usually suffer for want of food, and thus they did not rebel or abandon the site for better prospects elsewhere. They could almost always be sure that a fish could be caught after the requisite forest nuts had been gathered, rubber trees tapped, or some other extractive activity carried out. Entrepreneurs assumed that nature supplied a superabundance of fish, put there so that peasants could feed themselves free of charge—that is, free of charge to the baron.

Subsistence fishing is still the principal source of animal protein for the rural communities of the Amazon lowlands. But the fiefdoms are no longer secure. Owing to new attitudes toward property rights, conflicts (as will be discussed later) have arisen.

Except for one or two species, the ecology of Amazonian fisheries was probably little changed by humans until the 1960s, when the urban population began to increase rapidly. Before that time most fish were taken for subsistence. The commercial fleets were relatively small and only operated close to urban centers. It was from subsistence fishing, however, that Amazonian commercial fisheries evolved and are still shaped to some extent today. Urban fishermen were recruited from rural peasant communities that, in effect, provided the knowledge of natural history and the skills needed to operate in the very complex and often dangerous environments of Amazonian rivers and floodplains.

The policy of Brazil's military government to encourage occupation of vast reaches of the Amazon valley by spurring economic development led to the explosive and unplanned growth, beginning in the 1960s, of Manaus, Belém, and other cities. Rhetoric concerning the fisheries at the time was borrowed from the regional myth of superabundance. During the 1960s and even into the 1970s the myth of superabundance was confirmed by experience. As fleets grew and expanded into tributaries that had up to then been little exploited, the fisheries did indeed seem limitless. Until the end of the 1970s, fish sold in urban areas remained much cheaper than either chicken or beef. A study of fish consumption in Manaus in the late 1970s demonstrated a very high per capita daily intake of protein, averaging more than 150 grams. This was far more than the 50 grams recommended by most scientific studies as the minimum daily allowance.

Amazon fishing technology changed radically beginning in the late 1960s. Fish harvests before then were achieved using a wide variety of gear, including nets, gigs, bow and arrow, harpoons, bombs, and pole and line. These techniques were still

Local versus export fisheries. *Small-scale fishermen and industrial operations compete in the Amazon estuary. Small-scale operators account for only a minor part of the catfish catch, as most of it is exported by industrial interests.*

widely used in the 1970s, but large seines and gillnets accounted for most of the commercial catch, though not of all species.

Seines, as will be discussed in more detail later, were used mostly in the channels to catch fish species that seasonally used the great rivers for migration. Piranhas and small spiny catfishes discouraged the intensive use of seines in floodplain waters. There, gillnets proved to be much more versatile and economical; by the 1980s they had become standard equipment for urban, rural, and peasant fishermen. Gillnets were enthusiastically adopted, owing to the great dependence of Amazonians on fish and the realization that this gear was far easier to use and more effective for most species than were traditional methods. No other technological innovation in the history of the Amazon diffused as far and wide as gillnets in such a short period of time.

The estuarine fisheries near the mouth of the great Amazon River were exploited by small-scale operations until the 1970s. Then the Brazilian government began to help finance an expanding fleet, more sophisticated equipment, and refrigeration plants. Unlike the fisheries of inland waters, those of the estuary are now dominated by an industrial fleet aimed at exports to the national and international markets.

NATURAL HISTORY OF THE ESTUARY AND LOWER AMAZON FISHERIES

The low relief through which the Amazon River flows means that the Atlantic tides work their effects as much as a thousand kilometers upstream. But daily tidal cycles that bring noticeable fluctuations in the river level, thereby flooding forests and sustaining vast mudflats, are found only further downstream. The Rio Xingu, four hundred kilometers from the coast, thus marks the beginning of what is commonly called the lower Amazon. (This term is not to be confused with "lowland Amazonia" or "Amazon lowlands," which refer to any part of the Amazon Basin lower than about two hundred meters in elevation.)

Although estuarine floodplains are inundated by the daily tides, their forest communities are very similar to those found farther upstream in the seasonally flooded areas. The huge discharge of the Amazon River supplies enough freshwater during most of the year so that the estuarine forests are not heavily affected by salty marine invasions. Although there is a seasonal invasion of salt water during the low-water season of the Amazon River, fresh water, which is lighter, floats and flows on top of this intrusion. The tidal forests are thus inundated mostly by fresh water. Mangroves take over only near the outer coast.

Fish populations in most of the Amazon River estuary are noticeably different from those upstream. There is reduced diversity but very high productivity of a few species, and especially the juveniles of some of the large migratory catfishes. The lower Amazon and its estuary offer the single most productive large fisheries in the Amazon Basin. Even so, estuarine fisheries show dangerous signs of overexploitation. It should be kept in mind, too, that the ecology of the lower Amazon is linked to inland waters because of upstream and downstream fish migrations. For catfishes in particular, the health of the estuarine fishery can be no better than the health of the inland spawning grounds on which these fish depend.

Fish, shrimp, and crab are all important in estuarine fisheries. Most of the crab catch of the Amazon estuary consists of one brackish-water species. This crustacean is rather small, as commercial crabs go, and not very attractive, and for this reason there is no export market for it. Despite the crab's hirsute appearance and diminutive size, it is sold live for local consumption. This crab is very popular in Belém and other estuarine markets close to brackish water. Crab fisheries in the southern part of the estuary have recently been expanding rapidly as more fishermen shift from the exploitation of fish to crustaceans. The increasing price of fishing nets, and the lower returns per unit of effort and expense for fishing equipment compared with crab traps, appears to be the main reason for this shift.

The shrimp fisheries associated with the Amazon estuary embrace the brackish water zone and the offshore marine area to about a hundred meters depth. Three species are regularly exploited. All three species breed in brackish water near mangroves. Adults migrate to marine waters in a zone extending a hundred kilometers east of the island of Marajó. Both juvenile and adult populations are exploited. Industrial boats capture mostly adults in the offshore area, whereas the smaller artisan boats exploit juveniles in brackish waters. Most of the adult catch is exported. Total annual shrimp exports are now greater than those of the piramutaba catfish. Although the shrimp fisheries seem robust today, overexploitation is possible—as these fisheries are almost totally uncontrolled.

The piramutaba catfish, a predator, is the single most important fish species that has been exported from the Amazon region since the 1970s. By 1977 about twenty-five thousand tons of the catfish were landed in the estuarine area. In the late 1980s it ranked third in the list of all fish species—including marine species—exported from Brazil. At the peak of the fisheries, at least 70 percent of the total piramutaba catch was exported. The remainder was consumed locally. The value of the piramutaba harvest reached $13 million by 1980, but it is now less than $3 million because of declining catches.

In the industrial fisheries—that is, operations that use company fleets—

Belém, the largest city in the Amazon Basin.
Commercial fisheries have operated out of Belém since the last century, but only recently have attempts been made to study them. Since the early 1970s most of the fish catch in the estuary has been exported, which is in stark contrast to the central Amazon region, where most of the catch is consumed locally.

piramutaba are captured mostly with trawl seines. Small-scale fishermen depend primarily on gillnets. Although piramutaba migrate and may move as far as three thousand kilometers upstream to spawn, most of the commercial catch of the species is taken in the estuary because of the concentration of industrial-sized boats and access to refrigeration plants. The fact that most of the young size classes of piramutaba are restricted to the estuary also means that there is a much higher biomass of the species in the lower Amazon than upriver.

Development of the piramutaba industrial fishery was stimulated by minimal export restrictions, government-backed financing, and tax breaks to entrepreneurs. Governmental policies to promote this export fishery were not, however, informed by an understanding of the ecology. No serious studies were undertaken. Today, overexploitation is a serious problem, and the state and federal governments seem incapable of wresting control.

The piramutaba fishery will surely continue to decline unless initiatives, legally enforced by the government or self-imposed by the fishing industry, are taken to prohibit or control exports. As well as limits on the catch, effective management will depend on locating and protecting the still-unknown spawning habitats of piramutaba, which may be largely in the western Amazon Basin. Spawning habitats may be 2,000 to 3,500 kilometers inland from the estuary, and protection of those waters would thus involve Brazilian, Peruvian, and Bolivian governments. If the spawning habitats were polluted, say, by oil spills, this would also probably destroy the most important estuarine fisheries.

The dourada, which is also a predator and a migratory catfish, is the second most important species exploited in the estuary. It has a similar life history to that of the piramutaba. Industrial and small-scale fishermen of the estuary judge that the dourada population as a whole has a lower biomass than the piramutaba, and fisheries statistics seem to bear this out.

The total Amazonian dourada catch is divided between the estuary and inland waters. The dourada has a more predictable upstream migration pattern than does the piramutaba, and it also seems to disperse more evenly in the muddy rivers, the habitats where both species are mainly captured upstream of the estuary. Every year large schools of dourada migrate upstream and out of the estuary, whereas the piramutaba is more sporadic. Even young size classes of dourada are now being exploited by small-scale fisheries and industrial operations in the estuary. Immature dourada are now common in the Belém and other regional markets of the estuary. The exploitation of juveniles in the estuary is probably the greatest threat to the viability of this fishery. If the estuarine populations are overexploited, then the inland fisheries for dourada will also be destroyed.

An estuarine fishery that illustrates how little is known about the Amazon fish fauna is that of yet another kind of catfish, the rock-bacu. The rock-bacu is a large armored catfish, whose tough plates and strong spines are matched by no other Amazonian fish. Until the early 1980s this fish was known to science only from a few museum specimens that were captured in inland waters. Meanwhile, the locals were happily eating it. As it turns out, the rock-bacu has long been an important food fish!

Small-scale fishing for rock-bacu. *The rock-bacu fishery is one of several that serve local markets. These fish move in and out of estuarine forests with the tides.*

The main population of this species is found in the estuary. There it is a relatively important subsistence and commercial fish, especially in the smaller towns between Belém and the ocean. The rock-bacu, unlike the two other catfishes already discussed, may be able to tolerate some brackish water. During the low-water period of the Amazon River, when the eastern estuary becomes far more brackish than usual, the rock-bacu is found in the same areas as marine catfishes, though its seeming tolerance for salt water may owe instead to its keeping to the upper, freshwater layers. The rock-bacu is adapted to move in and out of the tidal forests. In these flooded forests it feeds on fleshy fruits and leaves. Weir traps that line many estuarine islands and shores capture this catfish and many other species that migrate into the inundated forests at high tide. Adult rock-bacu also migrate upstream seasonally; the juveniles then have the estuarine forests to themselves. But this fish is of little importance to the inland fisheries.

Food chains that support all the major ocean fisheries throughout the world are based on plankton production. Plankton is much less important in the Amazon River system, however. The reason is twofold. Muddy water reduces light penetration needed for algae growth, and many of the tributaries have poor nutrient levels. The most important plankton-feeding commercial species in the Amazon are the mapará catfishes, which are not estuarine species. Maparás have little or no tolerance for brackish water, and they do not therefore live in the estuarine areas of high plankton production. Instead these catfishes frequent the huge mouth bays of the right-bank clearwater tributaries of the Amazon River. From east to west these are the Rio Tocantins, Rio Xingu, and Rio Tapajós. Although also found in all river types throughout the Amazon Basin, maparás are particularly abundant in the floodplain lakes along the lower Rio Amazonas. Maparás are well known in Amazonian cuisine for high oil content, though traditionally they have been considered a second-class food fish. Maparás are more popular food fishes in the lower Amazon region than elsewhere. For at least two decades fishing boats from Belém and Macapá have concentrated seasonally on maparás in the lower Amazon region. Some of these boats travel as far as a thousand kilometers upstream to reach appropriate floodplain lakes.

Construction of the Tucuruí dam appears to have largely destroyed the mapará fisheries of the Rio Tocantins. Before impoundment was effected in 1984, annual mapará catches from the lower Rio Tocantins were three hundred tons. Maparás migrated upstream in the Rio Tocantins to spawn somewhere below the first rapids. Because the dam was built at the rapids, it is unclear whether destruction of the mapará fishery owes to a loss of spawning habitat by physical blockage or to some chemical alteration of the water wrought by impoundment or perhaps simply to overfishing.

Croakers, which are also called drums in English, are probably the most diverse group of food fishes captured for the Belém market. The commercially most important species are those that live in the brackish waters of the estuary, though marine and freshwater representatives are common as well. Croakers are predators of fish and crustaceans and are captured with gillnets or on line and hook. At least a dozen species are sold, and some of them are prized as first-class food fishes. Little is yet known about the ecology or fisheries of these species in the Amazon estuary.

Within the Amazon drainage, mullet (family Mugilidae) are restricted to the estuarine area, where they are one of the most important (if not relished) fish groups consumed locally. In general they are considered second-class species because of their somewhat muddy taste. Japanese restaurants in Belém, however, have enhanced their reputation as a result of better preparation. As catch sizes of the more favored species of estuarine fishes decline, mullet are likely to increase in importance.

Overall, fish and shrimp are taken in large quantities by both industrial and small-scale operations in the estuary and lower reaches of the Amazon River. Even so, those fisheries no longer supply the bulk of animal protein used by the major city in the Amazonian estuary, Belém. This is because estuarine fisheries are now primarily export-oriented. Cattle, buffalo, and poultry raised locally and supplemented with imports from central and southern Brazil thus supply most of Belém's animal protein—and at a considerably higher price than would be the case if estuarine catches were not exported. Since the mid 1970s most of the fish and shrimp catch from the estuary has been exported. Estuarine catches have been seriously declining in recent years because of uncontrolled overexploitation by industrial operations.

FISH WEALTH ALONG THE GREAT RIVER

The Amazon lowlands, all generally under two hundred meters in elevation, form a huge teardrop-shaped area of more than two million square kilometers. Oriented along the axis of the Amazon River, this region stretches from just downstream of Santarém in the Brazilian state of Pará to just beyond Iquitos in Peru. The species of food fishes of the Amazon lowlands are pretty much the same throughout this enormous area, and migratory patterns too are similar. To the east (estuarine) or west (uplands) of this region, migratory patterns and food fish species begin to change, though there is still much overlap. Some of the large catfishes that use the estuary, for example, are also found throughout the Amazon lowlands.

Although the diversity of fishes exploited for food in the Amazon lowlands is very high, several distinct types of fisheries account for most of the catch. In an ecological sense, fishermen are predators at the top of the food chain. Perhaps the most enlightening way to view the various fisheries, then, is to see where they fit into the overall food chain.

For fisheries in general, the lower a species is on the food chain the less chance that it will possess the flavor of a first-class food fish. Thus predators are usually considered to be the best-tasting fishes. The Amazon provides a striking and extremely important exception to the rule. A large number of fish groups are considered first-class, even though they feed at the bottom of the food chain.

This anomaly is created by the flooded forests. Those fine-tasting fishes are feeding heavily on fruits and seeds that fall out of the floodplain trees during the annual floods. Low-on-the-food-chain fishes that migrate into the flooded forests thus taste great. Perhaps the most extraordinary of these fruit- and seed-eating fishes is the tambaqui, which can grow to more than a meter and weigh thirty kilograms. This species has long been an important food species in the Amazon Basin. Alfred Russel Wallace and other naturalists praised its taste in the last century. At the end of the nineteenth century, during the full swing of the rubber boom, José Veríssimo—then the most astute observer of Amazonian fishing—reported that the tambaqui was captured in such large quantities for the Manaus market that the unsold excess was fed to prisoners. Today's prisoners could only dream of being served this delicious fish.

When the Manaus fisheries began to expand rapidly in the late 1960s, the tambaqui and a related species, the pirapitinga, were the main focus of much of the commercial fishing effort. Introduction of gillnets and seines ensured that these big fish could be captured throughout the year. Large seines were used to exploit schools encountered in the river channels either during spawning or dispersal migrations. Gillnets allowed fishermen to capture relatively large quantities of these fruit-eating species when they seasonally moved into the flooded forests.

During the 1970s the tambaqui became the single most important food fish species in the lowlands. For the Manaus market, which was and is by far the largest in central Amazonia, the tambaqui alone accounted for nearly half of the total catch sold in the mid 1970s. Since that time it has continued to be the single most important food species. Now, however, juvenile fish are heavily exploited, whereas in the 1970s most of the catch consisted of adults. Even as far upstream as Porto Velho, Rondônia, on the Rio Madeira, the tambaqui accounted for around a quarter of the total catch sold in the latter half of the 1980s.

In the span of just two decades adult tambaqui have been transformed from a relatively cheap to a very expensive food fish. Its large size and excellent flavor, derived from fruit and seed eating in flooded forests, makes the tambaqui one of the most popular food fishes in the Amazon Basin. Today large individuals are sold at prices equivalent to or surpassing those of choice cuts of beef, and they are a luxury item for the local upper class, restaurants, and export markets.

Management of the central Amazon tambaqui fishery poses both ecological and social problems that have larger implications. The foremost problem is the huge area embraced by the life history of the species. Beginning in the 1970s, the Manaus fishing fleet expanded in a mostly westward direction that included a large part of the tambaqui's distribution. At the time, the large pirapitinga was also one of the most abundant and heavily exploited food fishes, but within a decade it was nearing commercial extinction except for juveniles captured in the floodplain lakes fed by muddy rivers. Because no accurate fisheries statistics or studies were made in the early 1970s, when the species was most heavily exploited, it is unclear why the pirapitinga fishery declined so rapidly.

The most serious problem jeopardizing the recovery and eventual sustainability of the tambaqui fishery is overexploitation of juvenile fish. Juvenile fish are taken primarily in the floodplain lakes along the Amazon River and some of its muddy tributaries. Juveniles weighing only one to three kilograms are exploited more heavily each year, and there are no serious programs in place to protect them. This problem must be viewed on a regional scale. Juvenile tambaqui are one of the most important

Fishing with gillnets. *The large-scale introduction of gillnets in the 1970s was the single most important factor in the modern evolution of central Amazon fisheries. The gillnet replaced many traditional fishing methods, and this highly effective gear is now standard equipment in both subsistence and commercial fisheries.*

Fishing with castnets.
The castnet, which takes considerable skill to throw effectively, is one of the most common methods used in floodplain fisheries.

food fishes of the eastern part of the Amazon lowlands centering on the Santarém area. But few adult tambaqui are now found in the vicinity of Santarém. This absence is most likely due to the heavy deforestation of the floodplains that has taken place in the last two decades.

Adult tambaqui require large areas of flooded forest to find the fruits and seeds on which they depend. Since juvenile tambaqui are still relatively abundant in the floodplain nursery lakes, it seems safe to assume that these fish are recruited from upstream populations, most of which are probably above the mouth of the Rio Madeira, that is, in the area where adult populations of any size are still found. The recruitment takes place soon after the fish are born in the main river channel, at which time they are carried or swim downstream. Because floodplain populations, once reaching adult or nearly adult sizes, quit their nursery habitats and migrate upstream, it is highly probable that much of the adult recruitment in western Amazonia is made up of fish whose nurseries stretch along the middle Rio Amazonas to downstream of Santarém. This means that the overexploitation of young fish in the nursery habitats along the Rio Amazonas could also severely harm the adult tambaqui fisheries to the west. Likewise, destruction of adult populations in the western Amazon Basin could eventually eliminate the recruitment of young tambaqui into the nursery habitats along the Rio Amazonas.

Mud feeder is a term commonly applied to fishes that feed on various kinds of detritus, which is decomposing organic matter mixed with sand, clay, and tiny invertebrate animals. The term usually connotes an inferior food fish because of its muddy taste. The Amazon Basin has a great variety of detritus-feeding food fishes, most of which are less than 40 centimeters in length. In the Amazon, as in most places of the world, the so-called mud feeders are considered second-class food fishes. What mud feeders lack in flavor, however, they make up in numbers. Because these fishes feed low on the food chain, there is a bounty of food for them, and this largely explains their abundance.

Some of the most peculiar detritus feeders in the Amazon are the jaraqui. Jaraqui, of which there are several species, may look like carp, but they taste nothing like carp. They are generally considered to be the best-tasting detritus feeders in the Amazon. A look at their food habits offers an explanation. Although found in all water types, jaraqui have special adaptations to feed on the detritus attached to trees and other submerged substrates found in the flooded forests of nutrient-poor rivers. These fishes might well be called tree suckers. Able to make their mouths into suction cups by turning their thick and fleshy lips inside out, jaraqui vacuum minute food from submerged stems, limbs, and leaves.

In almost all markets of the central Amazon, jaraqui are among the ten most important food fishes. The largest and best-studied jaraqui fishery is that of the Rio Negro, which supplies the Manaus market. An intensive study carried out by Mauro Ribeiro of the Brazilian Institute of Geography and Statistics showed that two jaraqui species alone accounted for 90 percent of the commercial catch recorded from the lower Rio Negro in the early to mid 1980s. Following the tambaqui, jaraqui have been the second most important food fishes of the Manaus market. These catches come from both the Rio Negro and other rivers exploited by the Manaus fleet.

Tambaqui, the poster child for the flooded forest.
The Amazon is unusual in that predatory fishes are not the only species with tasty flesh. Fruit eaters such as the prized tambaqui—the most important food fish of the central Amazon—also appeal to the human palate. Because these fruit eaters are utterly dependent on flooded forests, the tambaqui is an appealing symbol for rallying support for environmental protection.

Jaraqui are usually the first of the migratory food fishes that fishermen encounter in large schools in the river channels during both spawning and dispersal movements. Entering the river channels four to six weeks earlier than the others, jaraqui are so well known that fishermen concentrate their fishing efforts on them. If other species were migrating at the same time, jaraqui populations would not be so heavily attacked. Most of the commercial catch is taken with seines, though gillnets and bombs are more important in some areas. Fishermen claim that bombing has altered the migratory behavior of jaraqui and that their movements are now less predictable.

Detritus feeders from the muddy rivers, such as the Amazon and Rio Madeira, do have a carplike taste. The most famous of these is the curimatá, a close relative of the jaraqui. Curimatá, which are of several species continent-wide, are the most important food fishes in nearly every major South American river basin west of the Andes. Curimatá are also important in the central Amazon, but never to the degree recorded elsewhere. In the central Amazon the one species of curimatá is restricted to muddy and clearwater rivers and their floodplains. It is rarely encountered in blackwater rivers. For example, it almost totally avoids the Rio Negro. Its main niche seems to be an ability to tap the detritus cycle in muddy river floodplains. Curimatá are considered second-class food fishes because of their muddy taste and bony flesh. They are captured with seines when migrating in the river channels and with gillnets when in floodplain waters. As the more favorite food fishes become rarer, the curimatá's importance to total catches is likely to increase, as are some of the other detritus feeders.

The Amazon has the greatest variety of predatory fish of any freshwater system in the world. Predatory fish account for three-fourths of the first-class food fish species sold in the Amazon. The vast majority of predatory fish, however, are too bony or too small to command first-class prices. A good example is the predatory piranhas. Piranhas are often among the most common fishes in floodplain waters, but they are only minimally exploited because of their low prices and the damage they do to fishing gear.

The most famous predatory food fish in the Amazon is the magnificent pirarucu. It is one of the largest freshwater species in the world. Pirarucu can grow to more than three meters and 150 kilograms, though fish this size are now very rare. This species is widely distributed in the Amazon Basin and in all its major river types. Fisheries accounts indicate that before about the 1960s the largest pirarucu populations were found in the floodplain lakes of the muddy rivers of the central Amazon. The species must once have been one of the key predators of these waters.

There is no evidence that the pirarucu was heavily fished by Indians before the arrival of Europeans. Dried scales and the rasplike tongue of the pirarucu were, however, used as implements and ornaments by native peoples. Without steel-tipped harpoons, the pirarucu was probably very difficult to kill. Times changed with the coming of the Portuguese. Portuguese colonists adopted pirarucu as a substitute for the traditional cod in their cuisine. By the end of the nineteenth century salted pirarucu had become one of the most important sources of animal protein in the Amazon Basin. It was even exported to northeastern Brazil.

The rubber boom stimulated pirarucu fisheries because of the need for a nonperishable protein source for urban centers and for latex collectors scattered along the many rivers. As urban tastes changed, beginning in about the 1970s, and as fish populations fell, fresh pirarucu began to supplant the salted product. In the late 1970s pirarucu accounted for about a tenth of the total catch sold in the main fish market in Manaus. Today the catch has been halved and is still declining despite some efforts to prohibit the sale of individuals less than one meter.

Pirarucu used to be killed mostly with harpoons thrown from small dugout canoes. Sturdy gillnets have now largely replaced the harpoon as the principal fishing gear, and fishermen in the last two decades have begun to capture juveniles that, even at 60 centimeters, are still larger than most of the other food fishes. Although the species has been virtually extinguished in some of the small river systems, it is not yet in danger of outright biological extinction. On the other hand, it is clear that most rivers have indeed been overexploited and that the only way to sustain the production of this species is to raise it in captivity. Fish farmers are, in fact, already experimenting with the pirarucu and achieving weight gains for captive fish of nine kilograms per year.

The most important predatory fishes exploited in Amazonian floodplain fisheries are in a group called (in English) the peacock bass. These fish, however, are not bass, but belong to the cichlid family. Bass are not native to South America. The English name is derived from their superficial resemblance to bass and their large, peacocklike eyespot on the tail. Nearly all species of peacock bass, owing to their great beauty and rich flavor, are considered first-class market fish. Taken as a whole, peacock bass are usually among the ten most common food fish groups in the markets of the central Amazon.

As predators peacock bass occupy a special niche, since they are highly specialized for actively hunting and chasing prey. Most other floodplain predators use the ambush technique. The abundance and productivity of peacock bass are in large part explained by the fact that they specialize on small fish. These predators are the most important food chain link between the large biomass of small, unmarketable fish and the commercial fisheries.

Although peacock bass are common and important subsistence food fishes in all river types of the Amazon lowlands, most of the commercial catch supplying the larger towns and cities is taken in the floodplain lakes of the muddy rivers. There are large-scale peacock bass fisheries in several of the Amazon River floodplain lakes near Manaus. One of the most famous is that of Lago dos Reis on the island of Careiro near the confluence of the Rio Negro and the Amazon River. Miguel Petrere of the University of São Paulo in Rio Claro calculated that in this island lake alone, more than a hundred tons of peacock bass were captured annually in the mid 1970s. Peacock bass are captured with a combination of traditional and modern gear. Gillnets are increasingly becoming more important, though the gig, a pronged spear used with a light source to stab the fish at night when they are resting near the surface, is still very important in these fisheries.

The Amazon boasts the most diverse catfish fauna in the world. Most of the large species are predators. Until the 1970s Amazonians in general held catfish in low esteem, despite the delicious flesh of some of the species. An abundance of other species and taboos against eating scaleless fish aborted any large-scale internal market for them. In the early 1970s large quantities of catfish were salted and exported from Brazil to Colombia. By the mid 1970s, with the opening of new highways from southern to northern Brazil, entrepreneurs began to export catfish captured in the central Amazon to São Paulo and other states. Several export companies also began to buy catfish for North American markets. The boom in the restaurant business in Manaus and

Pirarucu, one of the world's biggest freshwater fishes.
Pirarucu are known to attain lengths of three meters and weights of 150 kilograms, but now even the size shown here (enmeshed in a gillnet) is rare, owing to overfishing.

other Amazonian cities encouraged experimentation with catfish fillets, which were soon found to be highly acceptable. The diversification of the urban population and breakdown of taboos within the last decade contributed to the change of mind (and taste) about catfish. Just as important to the growing regard for catfish are the skyrocketing prices for many of the traditional first-class fish species—now too expensive for the average consumer.

Most of the predatory catfish catch taken in the central Amazon comes during their migrations in the river channels. There migrating dourada and piramutaba catfish are captured mostly with deep-water gillnets that are drifted downstream. Trotlines are employed to capture some of the other large species. Several species that are a half to a full meter long are common in floodplain waters, and they are regularly captured with gillnets.

More than fifty species of food fish form large schools and migrate in the muddy rivers during low water. Most of these species are second- or third-class food fishes, either because of their relatively small size or inferior flavor. In Brazil these migrations and the fisheries dependent on them are generally referred to as the *piracema.* In terms of total catch these are the most productive of all the fisheries upstream of the estuary because of the great concentration of fish in the river channels and the facility with which seines can be used.

Seines are large nets fifty to five hundred meters in length that can either be cinched into a bag when used in open water or simply pulled onto a beach in shallower habitats. No other gear employed in the inland waters of Amazonia can catch as much fish in a single effort. Seining is not, however, a favored technique everywhere in the Amazon lowlands. In the state of Pará strong sea breezes that blow up the Rio Amazonas create much choppier waters than what is usually encountered westward. It is thus much more difficult to spot schools and to use seines in the open river channels of Pará.

Depending on the intensity of low-water fish migrations, the *piracema* fisheries of the central Amazon can last two or three months. Seine fishermen exploiting low-water fish migrations have a tendency to oversupply urban markets, and there is much spoilage and discarding—especially if large quantities of first-class species arrive.

Although authorities and the general public tend to assume that exploitation of migratory spawning schools is the most predatory form of fishing, this is not always the case. Consider, for example, the migratory characins of the central Amazon that undergo two annual migrations—one for spawning (high water) and one for dispersal (low or high water, depending on the species). Even if the exploitation of spawning migrations were effectively prohibited, that might mean very little if the fishing effort targeting the same species during the low-water period is not also controlled. Most dual-migration species are actually easier to catch during the low-water period, when they are in relatively shallow river channels, than when they are driven by the urge to spawn at the beginning of the floods. In terms of decreasing the reproductive potential of a population it makes little difference whether fish are captured six months or six hours before they spawn.

Amazonian fishermen refer to highly mixed catches as "salads." These catches can include any of the species thus far mentioned, in addition to a great variety of others. Most highly mixed catches are taken in floodplain waters with a variety of traditional and modern gear. It is very difficult to track the relative importance of the more than fifty common "salad" species in any given floodplain area, and even taxonomists often have trouble accurately identifying all fishes sold in some markets. These highly mixed catches are becoming more important each year, as the preferred stocks decline. Unfortunately, mixed catches also include large quantities of juvenile fish of threatened stocks, as gillnet effectiveness depends on the size of the fish and not its age. Some species of increasing commercial importance in these "salad" catches are piranhas, many cichlids, the detritus-feeding characins, and juvenile tambaqui.

CITIES AND THE GIANT FISHBOWL

In terms of human geography the Amazon Basin is overwhelmingly urban, despite popular portrayals to the contrary. At least two-thirds of the region's inhabitants live in towns or cities. Manaus alone, with more than one million people, claims a greater population than all the small settlements and villages along the entire length of the Amazon River floodplain. Many of the towns along the main rivers have also witnessed explosive population growth since the 1960s or 1970s. Some have doubled or tripled in population in the last thirty years. This growth has influenced and changed the ecological and economic nature of fish markets.

Manaus Located where the Rio Negro meets the Rio Amazonas, Manaus is probably the largest freshwater fish market in the world that is based strictly on river fisheries. Somewhere between 30,000 and 50,000 tons of river fish are landed in Manaus each year. Manaus is also the most diverse freshwater fish market in the world. More than two hundred fish species could probably be recorded in the Manaus market in any given year, though only about a dozen of these make up most of catch. But recording fish offered for sale in Manaus is no easy task. By the mid 1970s slums began to surround the main city, and many new fish markets opened to serve them. The dozens of fish markets now found in Manaus make it very difficult to collect catch statistics.

As the population of Manaus grew rapidly in the 1970s, its fishing fleet increased to seven hundred or even nine hundred boats that had a ton or more of capacity. The largest boats could hold fifty tons of fish. During the 1970s the Manaus fleet expanded westward to exploit first-class species in the Rio Purus, Rio Juruá, and middle to upper Rio Solimões (Amazon River). The fleet also moved eastward to Itacoatiara on the Rio Amazonas and thence into the lower Rio Madeira. Fishing intensified greatly everywhere within a day's travel of Manaus.

As early as 1973 major conflicts erupted between commercial fishermen and the local communities living around floodplain lakes. The most publicized conflict was the so-called Janauacá fish war, marked by the deaths of twenty-three fishermen and peasants. Since then, a complex network of less hostile social interactions between urban fishermen and local floodplain communities has been nurtured. Notably, local residents are hired as stay-at-home fishermen by commercial

Fish catches arriving in Manaus during the low-water period. *Manaus is probably the largest fish market in the world based entirely on freshwater river catches.*

boats. Styrofoam boxes furnished with ice transported by Manaus boats allow floodplain peasants to store catches, which are periodically collected by commercial boats. In other cases, floodplain residents have purchased or built their own boats and begun to supply Manaus directly. A shift to gillnet technology has also helped floodplain residents make their catches commercially viable.

Since about 1980 the Manaus fishing fleet has had to confront rising fuel costs, declining stocks, and competition or hostility from other municipalities along the Amazon River. An additional factor that is usually overlooked in accounting for the woes of fishermen is floodplain deforestation, which has probably decreased food fish productivity. All these factors contributed to inflation of fish prices. The first-class species, which before about 1980 were at least occasionally purchased by all but the very poorest classes in Manaus, are now luxury foods. Second-class fish species are now about comparable to chicken in price. Per capita consumption of animal protein has thus probably fallen considerably since the mid 1970s.

Smaller cities along the middle Amazon Downstream from Manaus are at least a dozen small cities that range in population from 20,000 to 200,000. The fishing fleets of these urban areas and smaller towns usually are able to find sufficient fish within a day or two's reach of their local markets. The floodplains in this middle region of the Amazon are already heavily deforested, and livestock ranching now threatens the remaining forests. It is not yet known to what extent this deforestation has affected the overall fisheries of this region.

Itacoatiara, now with about 50,000 inhabitants, is the only small city on the Amazon River whose fisheries have been recorded in any detail during the last two decades. Itacoatiara is situated on the left bank of the Rio Amazonas. It is about 250 kilometers downstream of Manaus, to which it is linked by a paved highway. The city is nearly surrounded by the Amazon River floodplain, and it enjoys close proximity to large fluvial islands studded with enormous lakes. Most commercial fishing is done within a sixty-kilometer, or one-day, radius. This area is also exploited by the Manaus fleet, which concentrates on catfish in the main river and its side channels.

In the late 1970s the Itacoatiara fleet annually landed more than 3,500 tons taken from a 120-kilometer stretch of the Amazon River. Less than a decade after introduction, gillnets had become the most important gear used in the Itacoatiara fisheries, accounting for 43 percent of the annual catch, followed by seines at 32 percent. The other 25 percent of the catch was taken by traditional gear, such as gig, pole, trotline, harpoon, castnet, handline, and bow and arrow. In the late 1970s fish was at least three times cheaper than even the least expensive cuts of beef in Itacoatiara. The per capita protein consumption in the Itacoatiara area was then estimated to be about 104 grams per day, which is to say about twice the minimum recommended amount.

Further downstream is Santarém, which has a population of at least 200,000. Santarém is situated at the confluence of the Rio Amazonas and Rio Tapajós. It is thus midway between Manaus and Belém. The Amazon River floodplain is very broad in this region, but it is heavily deforested. When the Portuguese first settled this region in the eighteenth century, they established royal fisheries based mostly on manatee hunting. Though manatee hunting continues clandestinely, populations of the mammal have been so reduced that it is unimportant compared to fish catches.

Santarém began to grow rapidly in the 1970s in response to construction of the Curuá-una dam, the opening of a road that linked the city to southern Brazil, and that portion of the Amazon gold rush centering on the middle Rio Tapajós basin. Construction of the Cuiabá-Santarém and the Santarém-Curuá-una highways led to upland deforestation for cattle ranching and other agricultural activities. Degraded pastures on poor soils, however, soon turned the attention of cattle ranchers to the Amazon floodplain. There livestock could be grazed on natural grasses during the low-water period. During the 1980s this brought floodplain deforestation to the Santarém area at an accelerating pace. Cattle ranchers operating north of Santarém and north of the Rio Amazonas (an area that centers on the cities of Monte Alegre, Alenquer, and Óbidos) also contributed to floodplain deforestation. More than 90 percent of the floodplain forest in this region has now been destroyed. Adult tambaqui and other fruit- and seed-eating fishes, which were once important to the local fisheries, are now becoming rare, and this may be due more to habitat destruction than overexploitation.

Change also came to this region with the gold rush. During most of the 1980s the Rio Tapajós, a clearwater river under natural circumstances, was rendered muddy by gold-dredging operations. Any detrimental effect on food fishes of this increased turbidity is only speculative, since nearly all the important commercial species live or spawn in the highly muddy Amazon River as well. The fisheries for mapará catfish in the Rio Tapajós, nevertheless, became all but commercially extinct in the 1980s.

These fishes are plankton feeders, and increased turbidity may well have decreased primary production in the Rio Tapajós.

The principal fisheries supplying Santarém are based on mixed species catches from Amazon floodplain lakes within a fifty-kilometer radius of the city, along with seasonal takes of migratory characins moving down the Rio Tapajós to spawn in the Amazon River. The Santarém municipal government, with the help of the now-extinct fish agency SUDEPE, has very effectively prohibited the use of all types of seines in this region since the 1970s. This policy continues today with the help of IBAMA, Brazil's environmental agency. Gillnets account for most of the Santarém commercial fish catch. Ironically, though authorities are successful at prohibiting seines, the most intensive fish-bombing operations in all of the Amazon take place near the mouth of the Rio Tapajós. Rio Tapajós characin schools have been bombed continually for at least two decades, and fishermen have reported that this activity has altered migratory behavior.

Porto Velho A city of at least 200,000 and the capital of the state of Rondônia, Porto Velho is located on the right bank of the upper Rio Madeira. The series of cataracts just upstream mark the limits of navigation. The Porto Velho fishing fleet exploits a five-hundred-kilometer stretch between the cataracts and the mouth of the Rio Aripuanã, a downstream tributary. Porto Velho is linked to Cuiabá

Catfish catch at the Teotônio rapids near Port Velho. *These catfish were caught with castnets.*

in Mato Grosso State by a paved highway that, after its completion in 1983, spurred large-scale colonization and the explosive growth of Porto Velho. The gold rush of the 1980s in the upper Rio Madeira region also attracted huge numbers of miners.

Before about 1975 Porto Velho depended heavily on fish for animal protein. As the city grew, however, the Rio Madeira fisheries were insufficient to meet demand. Most of the Rio Madeira has a relatively small floodplain, compared to the Amazon, Rio Purus, and some other central Amazon rivers. Thus its fish productivity is considerably less. In the late 1970s Porto Velho fishermen expanded into fishing grounds considerably upstream—all the way to where the Rio Mamoré and Rio Guaporé enter the Rio Madeira. Fish caught that far away were transported to Porto Velho by truck from Guajará-Mirim at the Bolivian border. Manaus boats now also supply Porto Velho.

In the ten-year period from 1979 to 1989 the total annual catch taken by the Porto Velho fleet ranged between about 900 and 1,500 tons. No drastic changes in average annual landings punctuated that period, though catch per unit of effort was analyzed only in 1979. As a group the migratory characins are by far the most important food fishes exploited by the Porto Velho fleet. In the late 1980s, 40 to 50 percent of the commercial catch was fruit- and seed-eating fishes, and the tambaqui was the most important species.

During the 1980s the upper Rio Madeira region centering on the state of Rondônia suffered massive environmental changes. Large-scale deforestation, dam building, and gold mining were all then under way. Floodplain deforestation along the Rio Madeira was, however, minimal. The largest floodplain area near Porto Velho, called Cuniã, was designated a reserve (though peasants were allowed to remain in the area). The bulk of the deforestation occurred, rather, in the headwaters. Environmental damage in those tributaries was extensive. Increased turbidity was especially noticeable in the Rio Machado (Ji-paraná), which was caused by erosion from agricultural activity along the Cuiabá–Porto Velho highway and its feeder roads in the watershed. The Rio Jamari, the closest large tributary of the Rio Madeira to Porto Velho, was dammed in 1989.

Mining operations have brought sediments and mercury pollution to the right-bank tributaries upstream of Porto Velho, which were all clearwater rivers. By the mid 1980s gold dredging came to the Rio Madeira itself. Dredges now work the sediments from just below Porto Velho all the way to the Bolivian border. The Teotônio Cataract, a legendary fishery site of the Rio Madeira, became so crowded with dredges above and below the rapids that fishermen believed the seasonal fish migrations were being impaired. The main threat to upper Rio Madeira fisheries, however, is mercury contamination. Media reports on radio and television have warned of this danger, and thus fishermen and peasants have become watchful. Fishermen have, in fact, reported deformities in dourada catfish, for which they blame the miners.

The long-term sustainability of the upper Rio Madeira fisheries will depend on protecting the floodplain forests in the tributaries. Especially crucial are the right-bank tributaries below Porto Velho, to where most of the food fishes migrate during the high-water period. A large part of the Porto Velho catch comes from migratory schools moving in and out of these tributaries. In the short run, however, mercury contamination may be the greatest threat to Rio Madeira fisheries, as any major (and especially, substantiated) scare would probably destroy the market for these fish.

Belém Capital of the state of Pará, and with a population of at least 1.5 million and growing, Belém is the largest city in the entire Amazon Basin. The giant island of Marajó separates this estuarine city from the mouth of the Amazon River by about 250 kilometers. Belém is situated on the south side of Marajó Bay, into which flow the Rio Tocantins, Rio Guamá, and Rio Amazonas waters. The estuarine water at Belém, though tidally influenced, is fresh most of the year. It is nevertheless highly turbid because of Amazon River sediments transported through the Breves Channel west of Marajó. It is through the Breves Channel that the Amazon River and its inland fisheries are reached from Belém.

Because the Amazon estuary is highly productive, Belém supports industrial and small-scale fishing fleets. The former accounts for most of the commercial catch in the lower Amazon. Most of this commercial catch is exported. Since the 1970s about 8,000 to 30,000 tons of fish have been exported annually. Shrimp exports have been on a comparable scale. Belém's small-scale fisheries harvested annually about 3,000 to 5,000 tons of fish (destined mostly for local markets) during the 1980s. Industrial fishing operations are restricted to the estuary, whereas smaller boats based in Belém may travel a thousand kilometers up the Amazon River, as far as the Pará-Amazonas state border. Most of the fish eaten in Belém, however, is taken from Marajó Bay, the lower Rio Tocantins, lakes on the island of Marajó, and the lower Rio Xingu.

Marajó Bay supports the most important fisheries exploited by Belém's fleets. Industrial operations, with their larger and more seaworthy boats, also fish and shrimp oceanward, east of Marajó. This region is also exploited by the fishing fleets of smaller cities, such as Vigia and Colares on the south bank and Soure and Salvaterra on the north. The largest industrial fisheries target freshwater catfishes. Small-scale fisheries take a greater range of species. Species composition of catches changes seasonally, depending on water salinity. During the low-water season of the Amazon River, which is from August to December, marine water invades the estuary; freshwater species are displaced upstream. Brackish water and marine species move in, becoming temporarily dominant in the eastern part of Marajó Bay. During this period the species composition of the Belém market is quite diverse, with a combination of marine, brackish, and freshwater species. During the high-water period of the Amazon River, when the estuary is largely freshwater, various catfish species move eastward, and they then become the main species exploited by the Belém fleet.

Thus far, tidal flooded forests and mangroves on the islands and shores of Marajó Bay have remained largely intact. Logging operations are just beginning to have an impact. No serious industrial pollution has been reported, though a large bauxite-processing factory has been built on the island of Barcarena just west of

Belém. There has been concern about slurry being dumped into the bay. It is overexploitation of stocks, however, that poses the greatest present threat to Marajó Bay fisheries. Industrial operations are of most concern, as even today these fisheries are almost entirely unregulated.

On its south side, the island of Marajó has a large and highly productive freshwater lake region called Arari, which is sustained not by flooding of the Amazon River but by the heavy rains in this part of Brazil. Waters of the Arari spill into Marajó Bay by a small river that allows boat passage for most of the year. The Arari region has a long history of fisheries exploitation dating back to the seventeenth century, and it has supplied the Belém market on a regular basis for most of the twentieth century. Historically Arari was the easternmost large-scale fishery for the giant pirarucu. As in most of the Amazon Basin this species has been overexploited, and it is now only of minimal importance. Today the Arari fisheries supplying Belém encompass a relatively large number of species. Ecologically the inland Marajó fisheries are similar to those along much of the Amazon River.

OTHER WAYS OF EXPLOITING FISH

Aquarium Trade Fisheries The Amazon Basin is one of the major sources of wild fish for the world's aquarium trade. Ornamental fishes have been shipped out of the Amazon on a regular basis since the 1930s. Until the 1960s they were transported mostly by boat. Construction in 1967 of a first-class international airport to serve the free trade zone of Manaus was probably the single most important stimulus to the export of aquarium fishes. Manaus became linked to Miami by weekly flights that brought imported goods and transported aquarium fishes on the return. Manaus, located on the lower Rio Negro, was ideally located to serve as the main entrepôt for the Amazon's aquarium fish trade.

The most sought after species in the Amazon is the cardinal neon. It is found only in the middle and upper Rio Negro and its tributaries. Manaus is thus the gateway to get to it. Major aquarium trade fisheries also exist in the Leticia area of Colombia, near the Brazilian border, and around Iquitos in Peru. Yet Manaus is by far the largest exporter. Various species exported out of Manaus are taken in many rivers in the central Amazon, but the cardinal neon alone probably accounts for more than 80 percent of the total catch.

Government statistics on ornamental fish exports are unreliable and underreport actual take by several orders of magnitude. Entrepreneurs have a strong incentive to underreport—that way they can avoid taxes. The most reliable way to gather Amazon export statistics is probably not to be found in the Amazon. Rather, import records in Miami may provide the best data.

The aquarium trade fisheries of the Rio Negro have been built on an economic system that is essentially feudal because it depends on indentured peasants to catch fish. The very low price paid to aquarium trade fishermen allows traders in Manaus and other Amazonian cities to comfortably compete with fish farms in the United States and Asia. Peasant fishermen are advanced food and goods, which they then must repay with deliveries of aquarium fish. The average price paid, or discounted from debts, for cardinals captured by Rio Negro fishermen is about

Working the aquarium trade. *Aquarium-trade wild fisheries have operated in the Amazon since the 1930s, but it is only since major airports were built in the 1960s that large quantities of ornamental species could be exported.*

one (U.S.) dollar per thousand fish. By the time these fish are sold at retail, the markup is 500 to 1,000 percent. Another factor that allows exporters to compete in the international market is that the cardinal and several other popular Amazonian species, and especially some of the catfishes, are hard to breed in captivity.

Despite ideal conditions for ornamental fish farming, Amazonian entrepreneurs and government agencies have shown little interest in this potentially lucrative business, and researchers have largely ignored it. In light of the history of the aquarium business, it seems safe to assume that fish breeders in the southern United States or Asia will eventually begin to reproduce the cardinal neon. Other difficult-to-breed species will follow, and this could dampen interest in wild stocks and largely destroy the aquarium fisheries of the Rio Negro and other Amazonian rivers. It thus seems imperative that ornamental fish breeding and farming research begin immediately in the Amazon if aquarium species are to provide a sustainable industry for this region. Fish farming would also help prevent the overexploitation of precious species, such as the discus cichlids.

Food Fish Farming Interest in fish aquaculture in the Amazon Basin has been considerable, but farming of native fishes has only just begun. Owing to a lack of literature on native species, fish farmers tend to cultivate exotic species. Foremost are the African tilapia and the Asian carp. There are, as yet, no reports of farmed exotics, grown in ponds or dammed streams, invading the rivers. But judging from experiences elsewhere and the fact that introductions have not yet been studied in the Amazon, it is highly possible that some populations of these exotics have already become established.

Target of the aquarium trade. *The cardinal neon is the most important aquarium fish exported from the Amazon. Wild populations are now beginning to decline. The Amazon, with its vast waters and rich fish fauna, would be an ideal region for culturing aquarium species.*

The Amazon Basin has many species much more delicious, and which would command a higher market price, than either Tilapia or carp. Fortunately, fish farmers have begun to experiment with some of these. One of the most hopeful is the tambaqui, which has now been bred successfully in captivity by using hormonal injections. Carlos Araújo-Lima of the National Institute of Amazonian Research, in cooperation with the Amazon Rivers Program of the Rainforest Alliance, has recently chosen the tambaqui as a flagship fish with which to launch a project aimed at making fish culture information for native species available and accessible in the Amazon Basin.

The great diversity of the Amazon fish fauna suggests that a large gene pool is available for hybridization. Fish farming should be encouraged. It would be much less destructive of the floodplain than livestock ranching, and productivity of animal protein would also be greater. Furthermore, fish prices will continue to rise, at least for the first-class species, and thus aquaculture can eventually become competitive.

Most government-sponsored animal production research in the Amazon is, however, focused on cattle and buffalo. Large landowners, many of whom are politicians, exert a powerful influence. There is also a lot more information available on how to raise exotic cattle than on how to raise native fish. Even in the case of the floodplains, animal research institutes seem to be more interested in cattle and buffalo than in a mixture of aquaculture and crop farming—which would be far less destructive of the environment.

In almost all countries where fish hatcheries have been developed, the restocking of rivers, reservoirs, and streams has become common, though admittedly not always successful. Several North American fisheries, for example, depend on restocking. Brazilians are advancing rapidly in their ability to spawn fishes in captivity, and it will soon be technologically, though perhaps not economically, possible to restock Amazonian species that have been overfished.

Sport Fishing The sport fishing potential of the Amazon is quite high from an environmental point of view. With its huge numbers of great and small rivers, and with its rich fish fauna, the Amazon should be a piscatorial magnet for the sport industry. Amazonian countries, however, lack the infrastructure to develop this industry on a large scale. The Amazon region and its fishes are too poorly advertised to attract large numbers of fishermen. Furthermore, there are only a few competent entrepreneurs with decent boats and enough experience to guide successful fishing trips.

Santarém sponsors an international fishing contest each year for large-mouth fish, and especially for the extremely handsome and aggressively fighting peacock bass. Porto Velho in Rondônia has long sponsored a unique catfish fishing contest at the Teotônio Rapids just above the city. In the last seven or eight years, however, the Teotônio Rapids have been invaded by gold miners. It is no longer an aesthetic draw.

Within the last two decades tourism revenue attributed to sport fishing in the Amazon has, however, increased sharply. These visitors represent a valuable but

Tucunaré, the most popular sport fish group. *Inadequate infrastructure has hindered growth of a sport fish economy in the Amazon.*

to date untapped source of information about the potential and problems of developing sport fishing in the Amazon.

Overall, the fish resources of the Amazon are marked by a mix of over- and underexploitation. Prospects for enhancing their economic viability and long-term sustainability are clouded by the effects of floodplain deforestation along the Amazon River, potential contamination from mercury in some watersheds, and lack of basic research to rapidly develop fish aquaculture. All these problems, however, can be solved. The great potential of the Amazon's fish resource—truly, the principal treasure born of what we call "the floods of fortune"—suggests that fisheries should be given top priority in economic development plans.

CHAPTER SEVEN

A DELUGE OF USEFUL PLANTS

The variety of wild plants in the Amazonian floodplains that peasants exploit for food is impressive. But these subsistence uses are of minimal importance for the bulk of today's Amazonians who live in urban areas. Almost no native plants of the flooded forests and floating meadows have been developed for large-scale commercial sale.

Very few rainforest plants have ever been harvested for food on a large scale. The great majority do not produce fruits, seeds, or roots that can be eaten by humans, and most of those that do are not found in dense enough concentrations to be economically important. It is still unclear why so few plants were domesticated by native Amazonian peoples. A heavy reliance on riverine animal resources, combined with crops such as maize and manioc brought in from other environments, may have provided little incentive to domesticate more plants. There is also the possibility that the tragic loss of formerly dense populations of farming civilizations along the Amazon floodplain led to more than just the death of cultures. It probably spelled the demise of some domesticated plants adapted to the floodplain. A thorough ethnobotany of the Amazon floodplain has never been undertaken, but such an effort might well reveal species whose cultivation or semidomestication largely disappeared with the native agrarian cultures.

Although little deliberate selection appears to have taken place in the flooded forest plants that are still valued for subsistence by floodplain peasants today, many species are in a state of what ethnobotanists refer to as proto-domestication. Riverine peasants are continuing the process of experimentation with plants started by indigenous people thousands of years ago. Peasants are planting or tolerating useful plants on a small scale, both in isolated home sites and in villages.

Various fruits, seeds, and hearts-of-palm are the most important food plants collected on the floodplains. Because of rich soils and the wide variety of edible native species adapted to seasonal inundation, fruit farming has great potential on the Amazon floodplain. Rather than a nuisance, seasonal flooding can be an advantage. Nutrient-charged floodwaters of the region's great muddy rivers act as annual, or nearly annual, fertilizers. They also suppress weeds and pests. In the estuary, tidally powered elevations in river water perform the same function. Most floodplain plants fruit during the rainy, or high-water, season, although a few species also produce appreciable seed crops during the low-water period.

Potentially valuable species of fruiting floodplain trees are currently being ravaged by deforestation. Along the middle Amazon most of the edible fruit species of large size have already been destroyed by logging or clearing for cattle ranching. Palms are being cut down for their hearts-of-palm; few replanting and controlled harvest programs are in place. The many native species of fruit-eating fish relished by rural and urban inhabitants alike also suggest that floodplain orchards could serve as a food source for aquaculture operations. Several species of livestock other than cattle also fatten on floodplain fruits.

Overall, the inhabitants of the Amazon floodplain still tap a large number of wild plant resources from a variety of different habitats, though this activity is becoming more difficult as deforestation proceeds. Much oral knowledge of indigenous food plants has already been lost. Young people along most of the middle Amazon have never seen an intact floodplain forest. Traditional botanical knowledge will die with their grandparents and parents before it has even been properly recorded. If a broad biodiversity study of useful floodplain plants is not undertaken within the next few years, the rich oral history knowledge will be lost forever.

As early as the 1930s the French geographer Paul Le Cointe noted that more than seventy-five floodplain plants were used for a variety of purposes along the Amazon River. And it is almost certain that Le Cointe's list represented only a modest sample of what has been of use historically. Today many of the species he discussed are hard to find or have disappeared along the middle and lower Amazon, particularly the large trees that once grew in the formerly extensive floodplain forests.

The loss of floodplain woods in one area cannot be "compensated" simply by safeguarding forest elsewhere. Not only is the floodplain forest rich in species, but most of those species are not present in upland habitats. The composition of the floodplain forest, moreover, changes markedly along the course of the Amazon, and even species that are distributed from near the Andes to the Atlantic often show great genetic variability from one part of the range to the next.

PALMS

Palms can be put to a great variety of human uses. Not surprisingly, palms are known in the Amazon as the "trees of life." Early naturalists, notably Carl Friedrich Phillip von Martius, Alfred Russel Wallace, and Richard Spruce, were fascinated by Amazonian palms, and each wrote a monograph on this group.

Palms were essential plants for indigenous societies. Peasant communities along the Amazon still use them for many of the same food and construction purposes as did

Palms, the "trees of life."

Many species of floodplain trees could be cultivated for their fruits. The jauari palm shown here is the most common palm of the floodplain, and it grows well in disturbed sites. It is a prolific producer of fruit, which can be fed to small livestock or used in farming fruit-eating fish. Jauari fruits are not eaten by people, but the trees are sometimes exploited for hearts-of-palm.

the Amerindians. Because of the abundance of palms, however, there were few attempts to domesticate any of the species in the Amazon. A few palm species, such as the American oil palm, may have been introduced into the Amazon from elsewhere on the continent before the arrival of Europeans.

The Amazon has a rich palm flora, with at least twenty-two species in the Brazilian portion of the Amazon Basin alone. More than half occur on the Amazon floodplain, all of which have some value for the local population. Some palm species seem to naturally grow in often thick formations of thousands, if not tens of thousands, of individual trees. In the tropical rainforest this growing habit is unusual, as individuals of most tree species are sparsely distributed. The dense associations of these palms, then, suggests that they have special adaptations to avoid disease and predators, two of the main factors that cause a more dispersed distribution for most other plants in the rainforest.

The jauari is the most common palm seen along the Amazon River and most of its tributaries as well. It is highly adapted to take advantage of disturbed areas or to colonize beaches—even those with almost sterile sands. Its seeds are dispersed by water and fruit-eating fish, and thus they are able to reach suitable and newly disturbed habitats with each new flood. Hearts-of-palm are cut from jauari, and small extractive businesses for this wild product have opened in the Rio Negro area. Most of the middle Amazon, however, has been too deforested to support such an industry. No attempts have been made to plant jauari.

Jauari fruits are not eaten by humans, but they are commonly fed to livestock, especially pigs. Pigs can only benefit from the fleshy part of the fruit, however, as the large seed is too hard for them to crack. Several fish species feed on jauari fruit in the wild (and at least one species, the tambaqui, cracks open the seed). Jauari could undoubtedly be planted in large orchards for aquaculture operations, and both its pulp and seeds could be used.

The fruit of another palm of the Amazon floodplain, murumuru, is festooned with long, black spines. But this defense has not deterred humans. Murumuru fruits are gathered gingerly in the understory of floodplain forest for snacks. The fronds provide excellent cordage for making coarse, but durable, textiles. Mumbaca, a close relative of murumuru, is another spiny palm of the dark interior of floodplain forests. The stems of its fronds provide a strong wood for making bows, harpoon shafts, and other fishing equipment. Jacitara, a spiny climbing palm of the Amazon floodplain, is used to fashion the tipiti, a sleevelike device for squeezing the water out of manioc dough.

The African oil palm was introduced into Brazil during the colonial period. It is now especially common in the northeast Brazilian state of Bahia. In the last few decades, sizable plantations of African oil palm have also been established in several parts of the Brazilian and Peruvian Amazon. Various traits of the American oil palm have been tapped by plant breeders to improve the yields of oil palm plantations throughout the tropics. American oil palm has been crossed with its African cousin to improve the quality of the oil, impart disease resistance, and lower the stature of the palm for ease of harvesting.

American oil palm occurs sporadically along the Amazon from roughly the border between Amazonas and Pará westward to the Peruvian Amazon. This small palm is occasionally harvested for its oily fruits, which are fed to pigs, and to prepare folk remedies. The role of American oil palm in modern plantation agriculture illustrates the value of safeguarding plant communities as sources of genes for upgrading crops.

The most extensive palm groves in the Amazon are dominated by species of the genus *Mauritia*, known as buriti or miriti in Brazil and moriche in Spanish-speaking countries of South America. Their principal adaptation is an ability to withstand permanently waterlogged soils. Most other palms of the flooded forest have roots that require aerated (hence, dry) soil for at least part of each year. Miriti's supreme tolerance for water makes it especially common in true swamps of both the upper and lower Amazon and in many of the tributary systems. Along the Amazon River they are usually found on the banks of streams that enter the floodplain.

The savory and nutritious fruit of miriti is eaten throughout the Amazon. The fruit's fleshy pulp contains three times the vitamin A found in carrots and as much vitamin C as typically found in oranges. A century ago the great botanist Richard Spruce marveled that locals on the upper Rio Negro and Orinoco could be sustained for some time just on miriti. Miriti fruits—the size of golf balls—furnish many of the vitamins that might otherwise be lacking in a diet generally poor in vegetables. In a

Assaí palm and fruit.
The assaí is the most beautiful and economically valuable palm found on Amazonian floodplains. Despite the great demand for assaí hearts-of-palm and fruit for regional drinks and ice creams, little effort has been made to cultivate this species in plantations.

sense, Amazonian fruits have historically substituted for the salads and greens typical of diets in more temperate areas. Although miriti is appreciated throughout the Amazon, it is particularly popular in Peru. According to Christine Padoch, an ethnobotanist with the New York Botanical Garden, miriti is third most important fruit—after bananas and plantains—marketed in Iquitos, the largest town in the Peruvian Amazon.

The orange-colored pulp of miriti fruit is squeezed for its juice, which is used fresh, fermented into wine, or added as flavoring to ice cream. The miriti palm also produces a sugar-rich sap that is extracted by felling the tree and then bleeding it. This use is limited, however, and does not seem to have led to any major deforestation of palm groves. In some areas, though, people harvest miriti fruits not by climbing up the tall trunks but by cutting down the trees. Pith from rotten trees is used as compost or mulch in home gardens. The balsa-like wood of miriti is sometimes used for flotation devices, while the fronds are used to fashion tipitis. Beetle grubs are collected from fallen miriti trunks and are eaten as a delicacy in Iquitos and in parts of northwestern Amazonia and the adjacent grasslands of the Orinoco Basin.

Wild miriti palms annually produce at least sixty thousand tons of fruit along the Amazon, only a part of which is harvested. Clearly, the generous palm has untapped potential. With their enormous fruit crops, for example, these palms could be used more extensively to supply both wild and domesticated animals with food. Many fishes feed on miriti fruits that fall into the waters of flooded forests. Several species of game, such as brocket deer, peccaries, and tapir, also seek out the fallen fruits. Protection of wild miriti stands is thus a high priority for conservation. The species could also be planted, both to increase fishing and hunting yields, and to feed livestock. As in the case of jauari, pigs are fond of miriti fruits.

Overall, the miriti palm would be an ideal candidate for large-scale reforestation of backwater swamps overzealously cleared for cattle ranching or by loggers. Indeed, some of the dense stands of miriti along parts of the Amazon may trace back to plantings by indigenous peoples long ago. Even today this palm is occasionally planted on a small scale in home gardens and agroforestry plots in both the Peruvian and Brazilian Amazon.

The only other palm to rival miriti in importance for fruit production along the Amazon is the slender açaí. Its purple or light green fruits are gathered for local and regional markets, and hearts-of-palm are cut and canned for wider distribution. This graceful palm, widespread throughout the Amazon Basin, forms especially dense stands in the Amazon estuary. The açaí tolerates a wide range of environments, as long as it is near water. In the estuarine area, for example, populations are adapted to a daily influx of tidal (but nevertheless fresh) water. Upstream of the limit of tidal influences, near Óbidos, açaí may be flooded only once a year for a few weeks. This palm is also common along streams in upland forest.

Formerly dense stands appear to have been eliminated by deforestation along parts of the middle Amazon, such as in the Santarém area. Such loss of wild açaí is particularly worrisome because this palm provides an employment "chain reaction." At the base, farmers and extractivists rely on income gained from collecting the fruits. Others then take the marble-sized fruits to market. Buyers vie for the best lots, and platoons of distributors cart the fruits to sales outlets. Small kiosks and shops then spin the fruits in simple, cottage-industry machines to extract the pulp. The pulp is typically sold in plastic bags to be eaten fresh, particularly in the late afternoon. Another portion is frozen for the ice cream trade. Açaí juice is often sweetened with sugar and thickened with tapioca or manioc flour to form a porridge that is sometimes the last meal of the day. Some purists prefer the savory flavor of açaí "straight."

Collection of açaí fruits provides significant income for people living on alluvial islands of the lower Amazon. On Combu Island near Belém, for example, Anthony Anderson and Edviges Ioris of the Ford Foundation found that average family income exceeds $4,000 per year, most of it derived from selling açaí fruits. While not all residents of the Amazon floodplain are blessed with such dense stands of economically important trees, extractivism is clearly the mainstay for the riverine economy in some areas and contributes an important share of rural incomes in most other parts of the Amazon River.

Another reason that açaí has such a ripple effect in the regional economy is its value for heart-of-palm sales. Canneries focusing on açaí are located in the estuarine area of the Amazon. In addition to cannery jobs, others find work in harvesting and transporting the heart-of-palm. Some heart-of-palm from açaí is exported, but most is consumed by the well-to-do in urban areas of Brazil. If managed properly, açaí stands can provide heart-of-palm on a sustained basis. This is because the roots resprout when the trunk is felled. Selective pruning for heart-of-palm actually enhances fruit production of the grove as a whole. If, on the other hand, only the shoot is removed and the trunk is left standing, the palm dies. Some clandestine operators climb açaí to extract the palm hearts, thereby destroying the resource. Conflicts sometimes arise between those who depend on açaí for fruits, and those who harvest heart-of-palm.

Many of the palms already discussed yield valuable oils in their nuts. We predict that several floodplain palm species will eventually be grown on a commercial scale for their oil, both for fuel and cooking. With proper machinery, local communities could extract vegetable oils to generate electricity and to light lamps. At present, few rural inhabitants have access to electricity. Those that do are mostly hooked into diesel-powered generators, which are easily retrofitted for palm oil.

PLANTS FOR WEAVING, CAULKING, AND PADDING

Rural people in Amazonia often make their own twine from a variety of wild plants, including palms. Palms as fiber for hammocks has already been discussed, but there are many other uses for palm fibers. For example, leaflets of jauari palm are split and woven into hats, a favorite of fishermen who must cope with the midday sun. The leaf stems of jupati palm, which grows in the estuary area of the Amazon and also in some inland locations in Pará, are made into hats and small baskets.

Fiber needs are also met by other plants of the floodplain. Although many of the vines used for weaving come from upland areas, the floodplains do contain a variety of plants favored for baskets and sieves. In the Amazon estuary, locals fashion baskets from the stems of guarumã, an understory herb, to carry açaí fruits to market. Water-loving guarumã is also gathered in the forest to make sleevelike presses to squeeze the liquid from manioc dough.

Several floodplain plants produce spongy inner barks suitable for caulking, seed coats that are like cotton and can be used for stuffing, or stems that are suitable for padding. The bark of mamorana, a tree of the floodplain forest, is valued as caulking for canoes. The silky threads surrounding the seeds of the kapok, munguba, and mamorana trees are gathered to stuff mattresses and pillows. The Amazonian kapok tree was even introduced to Africa, Asia, and Polynesia for this purpose. For twine and padding, people also look to more open areas, such as floating meadows, lake margins, and emerging mud flats. The junco, a kind of rush, is collected from marshes to make pads that are placed underneath saddles. Thick mats of junco are commonly sold in stores and markets along the middle Amazon.

FRUITS AND NUTS

The Amazon Basin is famous for its delicious Brazil nut. The Brazil nut tree, however, is strictly an upland species. The rivers are important for the Brazil nut economy only as conduits for transportation. Huge cargoes of these nuts are often seen on rafts negotiating Amazon rivers.

Floodplain forests do have a diverse selection of species that produce large nuts. Most of these nuts, however, appear to be toxic to humans. Almost no research has yet been initiated to ascertain whether and how some of these nuts might be used.

The most promising floodplain species in the Brazil-nut family is the sapucaia nut.

Nuts from the floodplain.
The sapucaia tree produces the most locally sought-after nut of all species in the Brazil-nut family that grow in the floodplain. Unlike the exported species of Brazil nut, sapucaia seeds are not encased in a hard shell. When ripe, 15 to 25 seeds will drop out of the large, bowl-like capsules that bear them. Left: *nuts ripening. The white fleshy material is called an aril. Its function is to attract animals to disperse the seeds.* Right: *an empty capsule after the seeds have fallen.*

Sapucaia nuts are encased in a large, bowl-like capsule that hangs upside down. The tasty nuts are secured by a fleshy appendage (an aril to botanists) and are tamped inside the inverted bowl by a saucer-shaped lid. When the nuts are ripe, the lid falls off, at which time bats often abscond with the nuts to feed on the succulent arils. To subvert the bats or prevent the nuts from spilling on the ground, people frequently harvest the nuts just before they are fully ripe. Sapucaia capsules are sometimes used locally as flowerpots, and are also sold in urban centers to serve as ashtrays. Thick, spreading sapucaia trees provide generous shade, stabilize river banks, and are commonly planted or spared the ax around houses.

The fleshy fruit market in the Amazon has grown explosively in the last decade or so in response to urban population growth. Oranges, passion fruit, and papaya—all exotic breeds—are the leaders in upland agriculture, but there is increasing demand for some of the Amazon's native species, such as cupuaçu, a relative of cacao. Although citrus and papaya are planted on the higher parts of the floodplain, they do not produce as well there as in upland environments. From the standpoint of fruit-eating monkeys and fish, the floodplains truly are the orchards of the Amazon. Could they become so for humans?

The Amazon floodplain does have several species of wild passion fruit, and some of these are edible and could possibly be crossed with domesticated species to produce a marketable variety that could tolerate seasonal flooding. Moreover, dozens of species of fleshy fruits in many different plant families are already harvested by peasants and Indians in the floodplains. Several could undoubtedly be domesticated for urban markets, as they are already being harvested sporadically from wild populations where flooded forests still exist.

A floodplain fleshy fruit commonly seen in city markets is the hogplum. The grape-sized fruits are typically mixed with water and sugar to make a refreshing drink. Hogplum fruits are increasingly used in local ice creams and sherbets. Other acidic fruits gathered from floodplain forests to make juice or to eat raw include the uxi and several wild relatives of the cultivated guava.

Not all valuable fruits of the floodplain forests are consumed directly by people. Fruits of the cannonball tree, for example, are fed to pigs and chickens. This relative of the Brazil-nut tree produces fruit with brownish-pink and pungent pulp surrounding the seeds. The pulp is encased in a hard, round shell that must be split with a machete or ax before it is fed to livestock. Because of its usefulness as livestock feed, cannonball trees are often spared when people clear a home site along the banks of the Amazon.

To list all edible fleshy fruits found on the Amazon floodplain would tax the reader's patience. Suffice it to say that almost every imaginable shape and size is available. Fleshy pulp around the seeds of elongate capsules of some members of the legume family are much appreciated in the Amazon. Wild relatives of soursop probably could be crossed to produce sweeter types. Large shrubs in the coffee and quinine family, such as the genipap, produce enormous berries that can be used for juices and ice creams. A cheap fermented drink is prepared from fruits of genipap and sold in some neighborhood stores throughout the Brazilian Amazon as *licor de jenipapo*. Bacuri, another sporadic member of floodplain forest communities, produces creamy fruits that locals savor. If a bacuri seedling is spotted in a home garden, it is nurtured and kept free of weeds.

Uxi fruits from the natural orchards of the Amazon.
Traditionally, dozens of species have been gathered from floodplain forests, and several are now of commercial value. The uxi, popular for making beverages, is among them.

Fish baits.
Many colorful fruits have traditionally been used as bait for fruit-eating fish. The examples here represent four distinct families. Left to right: *Melastomataceae, Palmae, Passifloraceae, and Sapotaceae.*

Fruits and seeds of the Amazonian floodplains not only are used for food but are among the principal baits for catching fish. Indigenous people undoubtedly baited wood and bone hooks with a variety of fruits. Colonists adopted the practice, but substituted metal hooks. Today peasants continue this tradition.

Much, if not most, of the knowledge about the fruits and seeds that fishes eat was inherited from the Indians. Many Amazonian trees have indigenous names that reflect the animals, including fishes, that feed on them. Fruits and seeds that fish find attractive are valuable indicators of the type of genetic diversity that might be cultivated for animal feed and also for human consumption.

A complete list of the fruit and seed baits used would probably comprise two hundred species. Many palm fruits are popular baits, as fish are attracted to their fleshy parts. One way to capture jaraqui, for example, is to tie pieces of urucuri palm fruit to a fishing line a little above a cluster of four or five hooks. When the float indicates that jaraqui are rasping at the oily pulp of urucuri, the fisherman tugs vigorously on the line in the hope of snagging one or more of the unsuspecting fish. At high water, fishermen harvest the grapelike fruits of the slender marajá palm, which grows in small clusters in tranquil waters of the middle Amazon floodplain. The purple fruits of marajá are especially relished by tambaqui and pirapitinga.

Other fleshy fruits used as bait include wild guavas, soursops, and species in the coco plum family. Rubber tree seeds are a favorite of some of the larger fish species that have the strong dentition to crack them. Some of the fish bait species, such as the red-fruited socoró, are even planted or encouraged in home gardens. This list would be a good place to begin studying fruit and seed species that might be planted on a larger scale. As deforestation continues, however, much of this knowledge could be lost, especially if it is not soon recorded.

PLANTS FOR HOUSEHOLD USE, FUEL, AND MEDICINE

In North America and Europe, people generally visit a hardware store for construction supplies and items for their gardens. The same pattern prevails for the well-to-do in the larger towns in the Amazon. But most residents of the Amazon floodplain cannot afford such luxuries, so they must garner as many supplies as possible from their immediate environment. Men, women, and children paddle or walk into floodplain forests to gather wood, fiber for twine, and palm fronds for a wide variety of purposes, including construction of houses and nurseries for seedlings.

Even the simplest homes on the floodplain require some wood, if only for posts. A favored tree for such purposes is mulatto-wood, now increasingly rare on the floodplain to which it is confined. The piranha-tree is a preferred wood for stilts, essential for floodplain homes. The extremely hard wood of this tree resists termites and waterlogging. Piranha-trees have not yet been planted. This, despite the fact that floodplain residents are already having to turn to less favorable woods for posts and stilts.

Buoyant logs are used along the Amazon as rafts to float stores or corrals of livestock during the floods, and as docks for villages and other homesteads. It is now very difficult, however, to find useful logs floating down the Amazon River. And large standing trees suitable for this use are now rare on most of the floodplain. The few remaining giant kapok trees along the Amazon are being cut into rafts and for centerboards in the veneer trade. Floating cattle corrals are becoming rarer for lack of floating logs, such as the much favored assacu, but also because livestock are increasingly being taken off the floodplain during the high-water period and placed on upland pastures.

Timber is also in demand to build boats and canoes. Now with so many of the floodplain forests gone or logged-off along the most populated middle Amazon, craftsmen have been forced to buy durable wood from suppliers along the upper Amazon or on the uplands. In the middle Amazon, most boats and canoes are now built with itaúba, a giant in upland forests but increasingly rare because of overexploitation.

Another extremely important use of wood in the Amazon is for fuel. Most people in industrial countries give little thought to how their food is cooked—that is, what fuel cooks it. We tend to take electric or gas stoves for granted. Heat comes with the simple flick of a switch. The middle and upper classes in towns and cities in the Amazon rely almost exclusively on bottled propane or butane gas to cook their meals. An urban bias can blind us to the fact that most of humanity still depends on wood for cooking.

In rural areas of the Amazon, wood and charcoal are the primary cooking fuels. The search for enough wood to prepare meals is thus a daily chore for most residents

Fuelwood.
The gathering of firewood takes a heavy toll on floodplain trees, especially in areas already deforested. Floodplain inhabitants prefer to cook with propane, but it is often unavailable or too expensive.

of the floodplain. One can conceive of fuelwood shortages in dry regions, but it may be difficult to imagine a fuelwood crisis arising in the well-watered Amazon Basin. Yet as floodplain forests along the Amazon continue to shrink, fuelwood supplies are likely to emerge as a resource issue.

Unless research is begun now on indigenous fuelwood species for the floodplain, higher parts may eventually be covered by plantations of exotic eucalyptus. In most of the humid, lowland tropics throughout the world, few species currently rival eucalyptus in productivity for fuel or pulp. While plantations of eucalyptus are undoubtedly warranted in some large-scale operations, small holders might benefit from multipurpose trees that provide fuel, fiber, fruit, and other products. Many such candidates grow along the Amazon floodplain, and would help maintain some of the natural biodiversity of the region.

Fuelwood is gathered from floodplain forests, from driftwood left by receding floods, and to a lesser extent from home gardens. Not all woods have the same heating value; consequently, certain species, such as mulatto-wood and various kinds of ingá, are preferred for cooking and to prepare manioc flour. Few people bake bread along the Amazon, but many are accustomed to eating foods made from manioc flour. Preparation of otherwise-toxic manioc flour is time-consuming. Toasting one batch of the gritty flour on a metal griddle can take several hours. Slow-burning wood with a high heat value is thus favored. Many floodplain residents make manioc flour at low water to sell in the nearby towns and cities. Plentiful supplies of fuelwood are thus essential to their livelihoods. If fuelwood supplies falter on the Amazon floodplain, this important economic and subsistence activity is likely to come to a halt, with severe nutritional and socioeconomic repercussions.

Most people who live on the Amazon floodplain can ill afford prescription medicines, let alone the fees of private doctors. Public health care is generally available only in the larger towns and cities, entailing a time-consuming boat trip. Given the huge demand on understaffed public health services, long lines are common and patients may have to stay overnight in order to secure an appointment. In view of the expense and inconvenience of conventional medical care, many floodplain residents resort to folk remedies produced from wild and domesticated plants. Many of the plants used in folk remedies are cultivated around homes, but others are gathered from the wild. Most of the plants used in domestic treatments of ailments are gathered from floodplain forest, where the greatest diversity of plants is found.

Geographer Paul Le Cointe noted the medicinal uses of numerous floodplain plants in the early part of this century. How many of those plants are still used is not known. In some cases they may no longer be available because of deforestation. One can only imagine how extensively the indigenous peoples of the floodplain once used plants for treating illness. A few examples of how plants were used (and possibly are still used) in folk cures will illustrate the diversity of medicinal species.

Medicinal plants often have multiple uses. A salve for skin lesion, for example, is prepared from the sap of virola, a valuable timber tree now largely logged out from the middle and lower Amazon. Oil from the American oil palm is used to placate whooping cough. Several wild fruit trees also double as medicinal plants. Inner bark of hogplum, a favored fruit tree of floodplain forests, is used to make a tea for treating diarrhea. Roots of the versatile açaí palm are boiled, and the resulting tea is taken to treat swelling.

Diarrhea and dysentery are a common complaint of people without access to piped, potable water and efficient sewage systems. Not surprisingly then, several plants from the Amazon floodplain are employed to alleviate symptoms of undisciplined bowel movements. These include hogplum and the kapok tree. Intestinal worms are also prevalent in rural peoples without suitable toilet facilities, and several plants are used to purge these parasites, such as some of the figs. Species used to cleanse wounds include some legumes, the andiroba (famous for its oil), and others in various plant families. To lower fever, residents of the Amazon floodplain employ preparations made from virola, among other plants.

PLANTS THAT CAN THRIVE IN DEFORESTED FLOODPLAINS

Owing to the great amount of deforestation that has taken place on the floodplain, plant species adapted to colonize disturbed areas have done very well. These plants are invariably characterized by huge fruit or seed crops and impressive growth rates. These two characteristics would be extremely important if better uses could be found for these types of plants, as they could be easily cultivated.

There are several "weedlike" tree species, and each is adapted to a particular part of the floodplain. The most common woody species seen along the disturbed levees of the Amazon is a willow. Unfortunately its wood is too soft to be of much use in construction, and the fruit crop consists of tiny seeds adapted to dispersal by wind and water. Salicylic acid, or aspirin, has long been synthesized, and although willow trees are a rich source of this compound, they cannot compete with laboratory production. As deforestation of the Amazon floodplain proceeds, the willow tree is finding increasing value. Young willow trees are cut to stake tomatoes. Willow-tree stakes root readily when inserted in moist soil, and the fast-growing tree is of some use for living fences to protect home gardens from small livestock. The greatest role of the willow, however, is that it helps stabilize the soft river banks along levees, though this function is probably minimal compared to that of the intact rainforest that once existed in the same habitats.

One of the largest trees found in highly disturbed floodplain forests along the middle Amazon is the munguba, or silk-cotton tree. It does especially well in some of the lower parts of the floodplain that are deeply inundated each year. This fast-growing tree is protected from predators by vicious ants, so it is a good candidate for human cultivation. Its wood is too soft to be used for construction, but it makes a great compost. Most of the inner part of the munguba trunk will decay into a spongy mulchlike material only a year or so after the tree has been felled. This mulch is highly regarded for making compost to condition soil for vegetable growing. In fact, munguba mulch sells for at least the same price as cow manure, and it lasts a lot longer. It is also much better than sawdust, which is being used increasingly, for lack of adequate compost. As descendants of Japanese immigrants have shown in the Amazon, intensive vegetable plots using compost are more economical than extensively cultivated operations on poor soils. There will undoubtedly be increased demand for compost, and the munguba is a promising source.

It is worth remembering that today's weed may be tomorrow's crop. Around fields and along the margins of lakes, for example, one waist-high herb, camapu, produces fruits rich in vitamin C that are relished as snack food. Camapu is a relative of tomato and often grows abundantly in cleared areas. Camapu is a weed or a resource, depending on one's perspective. To modern, commercial vegetable growers on the Amazon floodplain, camapu can be a nuisance in large numbers; to others, the husked fruits of the volunteer plant are a welcome supplement to the diet.

Two near relatives of camapu have been domesticated in Central and South America. In Mexico, a species of camapu known locally as tomatillo or tomate verde has been domesticated to complement sauces. Aficionados of the Mexican breakfast dish huevos rancheros may not be aware that they are eating a former weed when they savor the green relish. One day, with changing tastes and market conditions, camapu could become a crop, switching places with tomatoes, which can also become weedy. Even if camapu never follows the same path as its close relative in Mexico, it serves to illustrate the value in closely scrutinizing all plants used by people in their environment.

CHAPTER EIGHT

FARMING THE FLOODPLAIN

In terms of potential land use the most striking geographic facts about the floodplain of the Amazon River are its large size and rich soils. Floods are often seen as the enemy of agriculture. In many river systems of the world, people have worked feverishly to hold back the waters. This option is not realistic for the Amazon floodplain because of the great seasonal fluctuation of water level and the extremely destructive effects any large dam would have.

Moreover, the annual floods of the Amazon river bestow a blessing on floodplain soils that owners of failed upland plots can only dream about. The floods bring a fresh flush of nutrients and topsoil. After the annual floods farmers enjoy a nearly weed-free planting surface, nearby water for irrigation if necessary, and cheap fluvial transportation to take produce to market.

Yet much of this potential is underutilized, despite the fact that agricultural yields are generally much higher on the rich Amazon floodplain than on the uplands. The Amazon floodplain could be a major producer of staple foods and high-value products such as fruits, nuts, timber, and vegetables. Furthermore, crop farming could be combined profitably with aquaculture and small livestock, including some indigenous species.

Tree farming has a particularly bright future on the Amazon floodplain. Yet no woody perennials are grown on a large scale along the Amazon today. Tree crops help secure river banks, and when several species are intercropped, they provide habitat for birds and other animals. Virtually no experimentation has been done with floodplain tree species and varieties. Perennial crop plants would need to be adapted to withstand or even take advantage of the floods.

Many of the land use systems now practiced on the floodplains are attempts to

implement upland kinds of farming during the nonflooded period. Little serious attention has been given to prospects for the large-scale use of livestock other than cattle and water buffalo. Nevertheless, floodplain peasant farmers are the informal caretakers of a largely unrecognized but important gene bank of potentially important crop plants. These they cultivate near their houses for personal consumption or occasional marketing. Some of these semidomesticates undoubtedly would have much greater value if more was known about them. This chapter explores how farmers are now using the Amazon River floodplain and the promise that a better use of biodiversity holds for both economic development and conservation.

CATTLE AND THE CLASH OF ECONOMY AND ECOLOGY

A lot of different kinds of small livestock are maintained on the floodplain during times of low water by farmers and ranchers alike. Some ranchers diversify their cattle

Small livestock.
Ducks, pigs, and sheep (shown here) and goats have been widely introduced to the Amazon River floodplain. They are especially favored by peasants who cannot afford cattle or water buffalo.

or water buffalo operations by adding sheep or goats. Both sheep and goats are raised primarily for meat, rather than wool or milk. Smallholders, on the other hand, concentrate more on pigs, ducks, and chickens.

Pigs, goats, sheep, muscovy ducks, turkey, guinea fowl, and chickens are often kept year-round on the floodplain if the family maintains its residence there at high water. During the floods small livestock are kept in raised or floating corrals. When waters recede, pigs may be confined to pens to keep them from wandering.

Smallholders on the Amazon floodplain gain more economically and nutritionally from small livestock than from cattle and water buffalo. Small livestock are able to fend for themselves much of the time and do not require significant clearing to create pasture. They are also easier to market, since they can be taken to town on any of the passenger boats that ply the Amazon and its numerous side channels. On the floodplain island of Combu near Belém, for example, pigs provide 10 percent of household income to smallholders during November, December, and January.

The greatest cause for environmental concern about livestock foraging on the floodplain is not, therefore, attributable to the small beasts and birds maintained by small farmers. The concern, rather, owes to increasing use of the floodplain for commercial ranching of cattle and water buffalo.

Cattle were brought to the New World when Columbus landed at Hispaniola on his second voyage in 1494. When they were introduced into Amazonia is unclear. Pockets of upland savanna and floodplain meadows were the only open landscapes suitable in the Amazon for cattle in the early colonial period. Forest clearing would not have led to productive grazing land, since none of the native grasses provide good pasture. Pasture grasses currently used in the Amazon and much of tropical America are derived from African species. These were not imported into Brazil until the eighteenth century.

On relatively undisturbed floodplains, grasslands are confined mostly to the margins of lakes and to new alluvial islands. During the flood stage, floating meadows form extensive mats along the margins of lakes and can clog slow-moving channels, but European breeds of cattle are ill equipped to exploit such seasonally available food. Spanish and Portuguese colonists were more familiar with the kind of ranching they could practice in terra firme environments. The early immigrants therefore concentrated their herds on upland savannas in Amazonia, which naturally occur on perhaps twenty million hectares, owing to high water tables or soils not favorable to trees. Records show that cattle were released on the broad savannas of Roraima, in the northern Amazon Basin, in 1787. A few cattle may have been kept by Portuguese along the middle and lower Amazon in the sixteenth and seventeenth centuries, but they were not then significant in the regional economy. Manaus was supplied by cattle from Roraima in the eighteenth and nineteenth centuries, suggesting that few if any herds were kept on the Amazon floodplain.

With the growth of towns and cities, particularly in the last few decades, cattle ranching has increased along the Amazon. The introduction of humped, zebu cattle from India in the late nineteenth century, especially the white Nelore breed, provided a boost to cattle production on the floodplain. Zebu cattle can tolerate heat and

Floating meadows as pasture.
Cattle are moved onto the floodplain before the end of the annual floods. At this time they feed in the water, having the effect of a giant mower. The destruction of floating meadows harms fish communities, especially the fry.

When the forage fails.
As the dry season progresses, floodplain grasses may become scarce. At this time of year cattle herds are commonly seen near the homes of floodplain cowboys.

humidity better than can breeds derived from Iberian races. Better adapted to riverine environments than the humpless criollo cattle brought in earlier, Nelore can wade into chest-high water to feed on floating grasses.

When rising waters become too deep, floodplain ranchers have three options. They can confine cattle to floating platforms (called *marombas*); they can pen them in small upland corrals (known locally as *caiçaras*) adjacent to the floodplain; or they can transfer them to upland pastures. The floating platform option is chosen mainly by small-scale ranchers and farmers with just a few head of cattle that can be stall-fed. Canoes laden with hand-cut floating grasses, such as pemembeca, had been a common sight on the floodplain of the middle Amazon during the rainy season. *Marombas* are less common now, however, owing to the severe flood in 1976 and to increased access to uplands created by new or improved roads. Small-scale farmers living along the interface of uplands and the Amazon floodplain often choose caiçaras for the few cattle they keep. As in the case of floating corrals, cattle are fed twice daily with floating grasses collected at dawn and at the close of day. Larger ranchers generally own, or rent, pasture on uplands at high water.

On a limited scale, raising cattle on the floodplain provides several benefits for local populations. Milk and cheese are important dietary supplements in rural stretches of the Amazon floodplain and in some of the towns and cities. One could argue that by using the natural meadows of the floodplain to fatten cattle for beef production, less of the upland forest needs to be cleared for pasture. The floating meadows, however, are important nursery habitats for fish, and this needs to be considered when evaluating floodplain ranching.

Water buffalo were brought to Brazil in the late nineteenth century. Water buffalo now exceed 800,000 head, more than half of which live along the middle and lower Amazon. Marajó Island at the mouth of the Amazon has been the mainstay for

When the floods come.
During peak flood season cattle must either be taken to the uplands or corralled in floating pens called marombas. *Their tenders must then cut grass by hand and transport it to the pens.*

Bigger problems with water buffalo. *Water buffalo are becoming more popular on the floodplain each year, as they are better adapted to the water than cattle and grow faster. But their ability to forage in flooded habitat makes them more of an environmental threat.*

water buffalo in the Amazon since the start. These hardy animals, raised for both meat and cheese production, have been introduced upstream only over the last several decades. Water buffalo first made their appearance in the Óbidos and Santarém areas in the 1950s at the initiative of Brazil's agricultural research service, EMBRAPA.

One indication of how quickly the water buffalo herd has grown is the estimate of a total of 305,000 head in all of Pará in 1975. The paucity of precise data on the numbers and location of water buffalo notwithstanding, it seems likely that the herd in Pará has grown by another several hundred thousand since 1975. Marajó Island alone now has close to a half million water buffalo scattered over the eastern portion of the island, a good part of which is periodically inundated. Meanwhile, water buffalo on the middle Amazon have become common enough to be shipped to markets as far afield as Belém, Macapá, and Manaus.

As cattle and water buffalo herds have increased dramatically over the last three decades, the area devoted to food crops has undoubtedly declined. Crops now can be grown only when fenced, or on ephemeral new islands where cattle or water buffalo have not yet gained a hoof-hold. Even fences may not keep hungry water buffalo away from succulent vegetables or maize. Ranchers are supposed to compensate farmers for any crop damage caused by free-range cattle and water buffalo, but they rarely do. Few floodplain ranchers fence their properties, and small farmers can only afford to do so for small vegetable and maize plots near their homes.

If the floodplain is to realize its potential as the next great agricultural frontier of Brazil, the contentious issue of crop growing versus ungulate grazing will have to be tackled. How the drama of increased water buffalo and cattle herds plays out will also have major implications for biodiversity, particularly in the remaining floodplain forests. In our view, the flora and fauna of the Amazon floodplain would have a brighter future if the balance tipped in favor of those who would rather farm than ranch. This would lead to more intensive farming practices and less destruction of forests and natural meadows.

Water buffalo ranching has been gaining momentum along the middle Amazon because of the greater adaptability of these animals than cattle to the floodplain. Water buffalo produce more milk and put on weight faster. The milk of water buffalo is 8 percent fat, compared to about 3 percent for cattle breeds commonly found in the Amazon. The higher fat content of water buffalo milk makes it preferred for cheese production. Water buffalo can begin to reproduce just two years after birth, instead of the three years normal for cattle. And last, but not least, water buffalo are less fussy about what they eat. On the downside, the broader diet of water buffalo results in greater destruction of vegetation.

The growth of water buffalo herds is not being prodded by artificial incentives. No major bank or tax credits are currently available for water buffalo or cattle raising. Banks do, of course, lend ranchers money to upgrade their operations, but at prevailing commercial rates. Given Brazil's high inflation during much of the 1980s and early 1990s, and accompanying adjustments of loan balances and interest payments, few ranchers are risking their enterprises as collateral for such loans. A common remark about bank loans for livestock and farming is that interest rates are counterproductive.

In the Amazon, dairy products are exempt from sales tax, but most floodplain ranchers manage for meat production, not milk. The limited market for dairy products in the Amazon, and the paucity of facilities for processing milk, have historically discouraged ranchers from concentrating on dairy operations. As habits change in the region, however, and as new dairies come on line, the incentive to raise water buffalo rather than cattle will only increase.

Water buffalo herds are likely to continue growing, at least in the vicinity of the middle Amazon. All ranchers interviewed in the vicinity of Santarém and Óbidos, and that currently ran at least some water buffalo, planned to increase the proportion of water buffalo on their properties. A new dairy plant, Amazomilk, was inaugurated on the outskirts of Santarém in 1993. This plant may be a harbinger of what is to come in other towns along the Amazon. Half of the anticipated milk production is expected to come from the floodplain—mostly from water buffalo. The other half will come from upland pastures. In the dry season, the Amazon floodplain is expected to supply some 70 percent of the milk. McDonald's in São Paulo has recently approached Amazomilk as a potential supplier of high-quality mozzarella from water buffalo milk for its chain of fast-food outlets in southern Brazil.

This first dairy for Santarém has linkages with other land use systems on the floodplain and uplands. Whey, a by-product of cheese making, will be sold to pig farmers in both the uplands and the floodplain. The dairy will obtain free scrap wood—mostly logged from the terra firme—from nearby sawmills to fuel its boiler that sterilizes the milk. Amazomilk plans to buy pulp of various regional fruits, such as cupuaçu, to flavor yogurt.

With a capacity of 30,000 liters a day, the Santarém dairy is modest by European and North American standards, but it is one of only a very few in the entire Brazilian Amazon. Most Amazonians, urban as well as rural, are not accustomed to drinking fresh milk, hence Amazomilk's emphasis on cheese production. Hot milk is commonly mixed with coffee at breakfast, but it is usually reconstituted from powdered milk. Many people in Pará are so used to mixing powdered milk in a blender that they will often eschew fresh milk when it is available. The sole dairy in Manaus, for example, makes powdered milk and then reconstitutes and sells it in one-liter plastic bags. Habits are changing in the region, however, in part owing to the influx of people from southern Brazil. Demand for fresh milk is thus likely to increase in the North.

Ever-increasing demand for meat is still the force driving the increases in cattle and water buffalo herds on the Amazon floodplain. More than two-thirds of Amazonian residents now reside in villages, towns, and cities. In 1994 Santarém had close to 300,000 inhabitants, while 1,300,000 people lived in Belém. The urban share of the regional population continues to grow as rural areas decline. Some parts of the floodplain now contain fewer people than they did a couple of decades ago, in part because of the unusually high flood of 1976 and the collapse of jute prices. Even if per capita consumption of beef declines in the future (as people become more concerned about the hazards of eating cholesterol-rich foods), demand for beef is likely to remain strong because of population growth. Water buffalo meat is usually sold as "beef" in markets, as few consumers can see or taste the difference.

The driving forces behind cattle and water buffalo ranching on the floodplain are similar to those operating on the uplands. Even though fiscal incentives for creating cattle pasture on the uplands were removed in the 1980s, pasture remains one of the dominant uses of land on terra firme. Most of the existing operations got their start from large investments made possible by the old system. This includes many of the upland pasture areas where floodplain livestock are taken during the floods. Low labor requirements, ease of transport to markets, a ready source of cash, and year-round demand for beef make cattle raising a relatively safe investment in any area accessible by boat or road.

One barometer of the economic viability of cattle raising is that banks, both commercial and state-owned, are reluctant to provide loans to the agricultural sector in the Brazilian Amazon except for enterprises involving cattle ranching or major crop plants such as oranges, black pepper, and coffee. Even farmers of subsistence crops find it attractive to keep a few head of cattle. Both on the floodplain and uplands, small farmers see cattle as economic security to carry them through crop failures and whose value will not be diminished by inflation.

How the small operators are managing their cattle in conjunction with their crops could provide insights on ways to resolve the conflict between raising crops and cattle ranching. With the decline of jute as a cash crop, small farmers are increasingly blending animal husbandry with agriculture and fishing. The mix of farming, livestock raising, and fishing varies. Some small-scale operators have essentially abandoned agriculture to concentrate on cattle and fishing.

The fact that small farmers on the floodplain tend to raise a few head of cattle in addition to crops complicates the issue for those who would argue that large livestock should be discouraged or even banned on the floodplain. Vegetable growers rely on cattle manure to maintain high yields. Planting beds are often raised and made of "cured" cattle dung. The only other nutrient supplement is silt, casually obtained from the muddy Amazon when the vegetables are watered.

Not all small farmers on the Amazon floodplain would therefore favor eliminating cattle along the river, should that be desired for purely ecological ends. Some large ranchers might be persuaded to reduce or eliminate herds after compensation, but reaching numerous small and medium-scale operators with subsidies to phase out cattle would be daunting. Subsidies would require funds from already overburdened public purses, and would probably not work anyway.

Cattle and water buffalo are thus here to stay on the Amazon floodplain, at least for the foreseeable future. Faced with that reality, improving the productivity of livestock and existing pastures might be the most practical way to alleviate pressure on, and thus ecological damage to, the remaining floodplain forests. Consider, too, that if cattle were prohibited or greatly cut back on the floodplain, upland deforestation would probably worsen as ranchers compensated for the loss of lush pastures at low water. There is thus no easy, cheap, and clean solution to the problems caused by a growing cattle industry in the Amazonian floodplain. Instead, the ecological damage caused by cattle and water buffalo will surely continue until alternative solutions are found. As mentioned in chapter 6, fish culture offers a realistic and environmentally friendly alternative to large-scale cattle ranching on the floodplain, and it needs to be promoted in the same manner that government agencies have championed bovines.

Fortunately, few if any small farmers appear to have opted for water buffalo, the most destructive livestock with regard to forests, floating meadows, and crops. Some small- and medium-sized farmers are, however, likely to switch to water buffalo after they observe their performance on large ranches. Water buffalo are integrated with the life of small farms in other parts of the world, such as in many parts of Southeast Asia and southern India. It is not unreasonable to expect a similar association to evolve in the Amazon.

In contrast to upland pastures, which are composed mostly of deliberately planted African species, floodplains are dominated by native grasses. Some of the floodplain grasses continue to elongate as the waters rise, thereby providing food for wading livestock. Natural colonization and seed dispersal are sufficient to ensure new pastures each year with the coming of the floods. As mentioned earlier, floodplain grass communities have increased greatly in the past two decades because of deforestation. Livestock densities in some floodplain areas have become so high, however, that the grasses are grazed to the ground even before the onset of the dry season. Along with grasses, other herbaceous plants, such as water hyacinths, are selectively browsed. A combination of cattle, buffalo, sheep, pigs, and goats—each with its own food preferences—leads to the complete elimination of floodplain meadows in some areas. The long-term ecological effects of these livestock on floodplain herbaceous communities warrants concern.

The destruction of floating meadows affects more than just the herbaceous plants. These habitats provide important nurseries and shelter for many fish species and for a wide variety of invertebrates. Livestock grazing in the water not only destroys the floating meadows; the large quantity of feces deposited, along with the plant material that has been uprooted or broken, accelerates decomposition of organic matter and can lead to oxygen depletion. Fish and other organisms must then flee these waters, or perish.

Soil compaction and trampling by livestock, and especially by buffalo and cattle, are serious threats to some floodplain herbaceous plant communities. A striking example of this is the current plight of the giant Amazon water lily, which is confined to muddy rivers in the Amazon Basin. As the world's largest lily this denizen of the floodplain is one of the signature plants of the Amazon. It is a natural wonder and a draw for tourists. Modest floodplain deforestation appears to have at first increased populations of giant water lily by creating more open habitats. Cattle seemed to have had little effect on these plants. The introduction of water buffalo, however, has dramatically changed the prospects for the giant Amazon water lily. Water buffalo exact a heavy toll on these tabletop-sized, yet delicate, plants during the low-water period when the animals trundle through floodplain pools. In years with an exceptionally long and dry low-water season, such as in 1991, the giant lilies are virtually obliterated in areas frequented by water buffalo.

CULTIVATING VEGETABLES AND CEREALS

The portion of the Amazon floodplain in crops is a fraction of the area devoted to pasture for cattle and water buffalo. Cultivated gardens and fields are not, therefore, a major cause of floodplain deforestation, but there are other environmental concerns.

The growing cities in the Amazon are being fed largely from farms in other parts of Brazil and abroad. Bread is a significant source of protein for the urban population in Amazonia. Rice, another important staple consumed in Amazonia, is not produced locally on a large scale, but in central Brazil.

The Amazon floodplain offers a number of features favorable to intensive food production. The annual floods deposit a fresh layer of alluvium that rejuvenates the soil. In contrast, upland areas of forest or second growth must be cut to make way for planted fields and then burned to provide ash fertilizer. When yields inevitably decline, slash-and-burn farmers must repeat the cycle. Floodplain farmers, on the other

The best soils in the Amazon.
In the floodplains of sediment-rich rivers are found rich alluvial soils. The annual floods replenish nutrients, and they are easily tilled.

hand, can cultivate the same plot every year without resorting to fertilizers to shore up yields.

Another advantage of the floodplain for annual crops is that many weeds are destroyed at high water, and the farmers are presented with a relatively clean planting surface each time. Some native grasses, such as muri, are uprooted and burned, and the resulting ash further enriches the soil. Grasses along the banks of the Amazon and margins of lakes are also burned by small farmers to encourage nesting by side-necked turtles. Telltale tracks can then be easily followed to the eggs, a highly prized food.

Floodplain farmers can also take advantage of nearby water for irrigation. Although many crops, such as manioc and sweet potato, produce well without watering, yields of maize and vegetables increase when the soil is kept moist during the dry season. Finally, access to cheap water transportation aids the marketing of crops.

For the most part, crop production is not yet a significant cause of deforestation in the floodplains. A small section of floodplain forest may occasionally be felled for a plot of maize or manioc, but farming is largely carried out on already cleared land. During the heyday of jute, some floodplain forest was cut down, mostly in the 1950s and 1960s. Improved food production on the floodplain is thus unlikely to force further retreat of the forests.

The desire to increase crop production, while not as strong as the drive to increase cattle and water buffalo herds, is being spurred by two forces. First, urban populations are increasing in the region, therefore improving market opportunities. Second, after the collapse of jute prices, farmers are scrambling for other options. A move to more intensive farming methods is under way on the Amazon floodplain, particularly with market gardening and mechanized cereal production.

Vegetables have emerged as one of the most promising options for floodplain farmers after the collapse of jute. With the spectacular growth of cities, demand for a wide variety of vegetables—particularly tomato, lettuce, cabbage, cucumber, bell pepper, okra, and spring onion—has increased dramatically. Traditionally, people in Amazonia have not eaten many vegetables, but customs are changing with the rise of the middle class in urban areas and the influx of people from other regions. Salads are now commonly served in homes, restaurants, and in fast-food outlets in Amazonian cities.

With the paving of the Belém-Brasília highway and construction of the Transamazon Highway in the early 1970s, growers in southern Brazil initially benefited most from the growing appetite for vegetables in such towns as Belém, Santarém, and Manaus. Locally grown vegetables in the Brazilian Amazon had, until recently, come largely from intensively managed farms near the cities. Even so, those farms satisfied only a portion of demand. Imports from outside the floodplain were the dominant supply. Most of the locally grown vegetables consumed in Amazonia in the 1970s and 1980s probably came from uplands.

That picture is changing, at least in the middle Amazon. There, floodplain farmers are increasing their market share for vegetables in urban areas. For example, local growers on the floodplain now supply virtually all the tomatoes entering the markets of Santarém. Vegetables grown on the floodplain are produced in large part by small farmers with heavy reliance on family labor. Wage laborers, paid the equivalent of a dollar a day, including lunch, also find employment on small vegetable farms, particularly during the tomato harvest.

Vegetable growers are intensifying their production techniques. Vegetable plots are typically fenced to keep out cattle. The high return on vegetables apparently pays for the investment in fence, although cattle, and particularly water buffalo, still pose problems for vegetable growers in some areas. Some farmers have built raised platforms so that they can grow vegetables year-round, even during the floods. Many farmers have purchased diesel pumps to irrigate their vegetable plots. In some cases, several families have cooperated to buy and maintain portable irrigation pumps.

Vegetable farming on the floodplain has forged links with other land use systems. Cattle dung, obtained free from local ranches or purchased at a modest price, is either mixed with topsoil or used "pure" in vegetable beds. Cattle are usually corralled at night to reduce theft and to facilitate milking. Overnight droppings are easily bagged in polypropylene sacks for distribution to vegetable growers. Cattle are thus viewed ambivalently by vegetable growers. Constant vigilance is required to make sure fences are not breached by livestock. On the other hand, their manure fertilizes vegetables.

Farmers are also making use of sawdust from local mills. Sawdust is not a great nutrient, but it does suppress weeds and conserve soil moisture. Proliferation of sawmills in many Amazonian towns and cities is an outcome of the greater access to upland forests offered by pioneer roads and by the soaring prices for all types of hardwood for both domestic and international markets.

The spread of vegetable farming along the Amazon floodplain is undoubtedly improving rural incomes and employment opportunities. As well, it is improving the vitamin and fiber content of diets in urban areas. But market gardening also raises some ecological issues. Vegetable production, particularly in the wet tropics, is notorious for heavy use of insecticides and fungicides. For floodplain economies in which fisheries predominate, contamination of fish food chains would be dire indeed.

Little is known about which insecticides and fungicides are being used on the Amazon floodplain and the quantities involved. Farmers purchase pesticides from farm and livestock supply stores where they buy many of their vegetable seeds. Widespread illiteracy in rural Amazonia means that chemicals may not be mixed correctly and applied in appropriate doses. Irrigation allows two tomato crops to be grown on the floodplain, and if anything like the recommended doses are being used by vegetable growers, pesticides could pose a significant ecological and public health threat.

One way to reduce reliance on insecticides would be to deploy pest-resistant breeds. At the moment, farmers are purchasing seed produced in central and southern Brazil, or even imported from the United States and Europe; these breeds have not stood the test of Amazonia's pests. Pest pressure is less severe in the temperate climates where most vegetable varieties are developed. Vegetable growers in Amazonia would surely benefit from more research on varieties adapted to the onslaught of diseases and pests characteristic of the humid tropics.

Some ranchers along the middle and lower Amazon have begun experimenting

Coping with the floods.
*Raised gardens are one way to keep farming through the floods. Small gardens (*right*) are especially popular for growing green onions (shown here) and an assortment of medicinal plants. Large, platform methods of gardening (*above*) supply urban centers with onions, lettuce, chicory, cilantro, and other valued foods.*

with more intensive techniques for growing maize and rice. This move to mechanize farming on the floodplain is being driven by three main forces. First, local and regional investors are looking for other options, now that gold-mining profits are down. Second, investors from central and southern Brazil are buying properties on the floodplain and seeking ways to develop them. Finally, local market prospects for maize have improved because some urban centers have sizable poultry farms. Increased demand has drawn locally and regionally owned landholdings into maize production.

A large poultry operation in Santarém illustrates the growing market for livestock feed in the Amazon. Brazil's leading airline, Varig, established a poultry-processing plant in town, as well as a modern chicken farm along a side road of the Santarém-Rurópolis highway. The chickens are marketed widely, from Manaus to Macapá. Almost all the maize purchased to feed the chickens comes from growers south of Amazonia, in Goiás and Mato Grosso. The maize is first hauled by road to Belém or Manaus, and then barged to Santarém.

Locally grown maize for the chicken operation in Santarém would have two economic advantages. First, the appreciable transportation costs of moving maize 1,500 to 2,500 kilometers from other parts of Brazil would be greatly reduced. Second, those maize growers and dealers demand payment in advance, rather than upon delivery at the final destination. Maize purchased in Mato Grosso and Goiás typically takes thirteen to fifteen days to reach its destination, during which time the money is "tied up." If owners could buy maize locally, chickens would be converting the cereal to meat within a day or two.

In response to the sizable market for maize in Santarém, some farmers and ranchers have begun experimenting with mechanized and irrigated maize production on the Amazon floodplain. For example, one 5,000-hectare ranch with 2,000 head of cattle on a floodplain island thirty kilometers downstream from Santarém set aside 70 hectares for maize production in 1992. The maize area was disked with a tractor and periodically irrigated with silty river water. The tractor and irrigation equipment were purchased with fiscal incentives provided by a financial foundation administered through the state-owned Banco da Amazônia.

Water is drawn from the river by a diesel-powered pump set on a mobile barge that is otherwise used to transport cattle on and off the floodplain. When the maize starts to mature, irrigation ceases. Small amounts of potassium fertilizer are applied, but no insecticides or herbicides are currently used. A medium-stature variety of maize was chosen because it withstands the strong summer breezes that blow up the Amazon. Yields were expected to be around five tons per hectare—about five times typical yields on Amazonian uplands. But the first harvest (in late 1992) produced only three tons per hectare, about the same as yields achieved by small farmers on the floodplain without mechanization, irrigation, or fertilizers.

Prolific weeds and improper irrigation apparently contributed to the disappointing yields on that ranch. Although the receding floods present a relatively clean planting surface, various species of weeds soon occupy the moist soil. Traditional maize varieties on the Amazon floodplain are tall, however, and tend to outpace weeds; but the ranch in question chose to plant a medium-stature breed. Another reason that floodplain farmers have traditionally preferred tall maize is that the cobs can still be harvested as the floodwaters rise.

To solve the weed problem, at least one large landholder near Santarém plans to use herbicides in his mechanized maize operation. Herbicides are commonly employed on modern farms in many parts of the world, often with minimal environmental impact. The heavy use of such chemicals on the Amazon floodplain, where there are many complex food chains, could wreak havoc in aquatic ecosystems, at least on a local scale. A better approach might be to assess the potential for intensive production of some of the traditional varieties of maize, or to select tall hybrids that can tolerate close spacing to shade out weeds.

Some market-oriented maize growers annually achieve six to eight tons per hectare. They do this by raising two crops. The first crop is irrigated during the dry season as the flood recedes in July and August. The second crop does not require watering, as it is planted at the onset of the rains in November or December. Irrigation in itself is unlikely to result in any environmental problems, such as the buildup of salts often found in arid regions. Periodic inundation should counteract any alkalinity. Furthermore, irrigation in the floodplains of muddy rivers may actually enhance soil fertility by depositing a fine layer of silt.

Investors are also experimenting with intensified rice production on the Amazon floodplain. The owner of a cereal-drying plant in Santarém mechanized eight hundred hectares of floodplain near Alenquer for rice production in 1993; his yields averaged six tons per hectare—about six times the norm for rice farmers on uplands. Despite these economically promising returns, it would be unwise to put all agricultural eggs in one basket. The ecological cost would be too great.

Several intrinsic factors are likely to promote a more complex, mosaic pattern of land use anyway. The three main obstacles to more intensive rice cultivation are the sloping terrain; the cost of diesel fuel for irrigation; and the uneven arrival of floodwaters. If floods come early, large-scale rice operators stand to lose a substantial investment. Floodplain farmers will undoubtedly continue to experiment with cost-effective ways to intensify rice production, as they have done along parts of the Rio Branco in Roraima. But the Amazon floodplain is unlikely to become a breadbasket.

In Amazonia, rice is still mostly an upland crop. There, yields average one ton per hectare, and only one cycle can be grown during the rainy season. In contrast, highly mechanized rice growers in California (who are also limited to just one crop per year) typically achieve yields of eight tons per hectare. In terraced parts of Southeast Asia, such as Java and the northern Philippines, small-scale rice farmers can do even better by planting three crops per year, each with an average yield of three tons per hectare. On the Amazon floodplain, rice is cultivated mostly on the lower Amazon using traditional, low-input management, and yields of three tons per hectare are reported.

An American entrepreneur made the boldest, albeit unsuccessful, attempt to intensify rice production on the Amazon floodplain. Daniel Ludwig, a billionaire who made a fortune in shipping (and who is now deceased), acquired more than a million hectares of land straddling the states of Pará and Amapá in the 1960s. His aim was to

Preparing manioc flour. *The main starch food of floodplain inhabitants is manioc. It is eaten mostly as flour mixed with other foods or sprinkled on top. Manioc flour is roasted in large pans over open fires. In areas that have been heavily deforested, such as the floodplain of the middle Amazon River, the fuelwood demands of manioc processing threaten the remaining trees.*

produce pulp from tree plantations. The Jari operation soon diversified into cattle and kaolin mining, and Ludwig was apparently intrigued with the agricultural potential of the Amazon floodplain along the southern fringe of this property.

Under Ludwig's leadership, the Jari operation commenced large-scale rice production on the lower Amazon floodplain in the 1970s. Parts of the floodplain were diked, and rice production was heavily mechanized, including combine harvesters and light aircraft for spraying. While yields were reportedly high, between five and ten tons per hectare, outlays were exorbitant for the market value of the product. The cost of operating the diesel pumps for controlling water levels was particularly onerous. The former Jari rice lands (under new ownership) are now being used experimentally to raise pirarucu fish.

MAKING THE MOST OF TRADITIONAL FARMING

Experience has shown that intensifying food production in Amazonia with purchased inputs, such as fertilizers and pesticides, is often not cost-effective. The low prices that staples command in the market endanger any food-producing operation that requires heavy investments. Much can be learned, however, about how traditional food-producing systems are changing on the floodplain in response to the growing market for staples in towns and cities. Some of the low-cost strategies employed by small farmers could probably be improved to increase yields without damaging the environment. In Roraima, for example, enterprising small- to medium-scale farmers have recently begun irrigating rice on the floodplain of the Rio Branco, although this river has much poorer soils than has the Amazon.

Government leaders who wish to encourage food production on the Amazon should avoid any romantic ideas about returning to traditional methods or simply promoting existing, low-input systems. Such static notions about premodern agriculture fail to consider the dynamism of all agricultural systems, including traditional ones. Rather, a fuller appreciation of the range of technologies and management strategies of traditional systems is needed as a precursor to any interventions.

Parts of traditional systems, and especially their role as gene banks for cultivated plants, could surely play a much wider role in enhanced food production. A better understanding of traditional food-producing systems would also help identify bottlenecks and suggest remedial measures. The marketing and price structure of foodstuffs in the Amazon also needs checking to see whether artificially low food prices discourage farmers from producing staples for sale.

Two traditional food crops of the floodplain that deserve particular attention are squashes and manioc root. Squashes produce abundant harvests on the floodplain, both for home use and the market. Breeds of floodplain squashes developed in the Amazon may have been imprudently overlooked by the rest of the world as an important source of genetic variability. A riot of colors, shapes, and textures of squashes graces vegetable markets throughout the middle and lower Amazon. Heirloom squash varieties maintained in the agricultural "backwaters" of the Amazon floodplain could harbor valuable genes and promising material for more intensive squash production along the Amazon—and elsewhere.

Diversity is also pronounced among manioc varieties cultivated on the floodplain. Both bitter and sweet forms are grown, as they are throughout the Amazon, but the varieties are mostly different from those encountered in the uplands. Upland varieties tend to take at least a year to mature, which is too long for the seasonally inundated floodplain. If a farmer lives along the margin between uplands and floodplain, the family may plant several varieties of manioc in both environments. If, on the other hand, the farm is a long way from the uplands, varieties exclusive to the floodplain are preferred. The separation of varieties in the different farming environments is striking and demonstrates the fine-tuning by farmers as they adapt crops to the ever-changing landscapes of the Amazon floodplain.

More than forty varieties of manioc are planted in the region of the middle Amazon. Only a few of these, however, are found on both the floodplain and the uplands. Even accounting for the possibility of double counting—as one variety may go by different names in different areas—farmers everywhere in the Amazon maintain a wide assortment of distinct varieties. Why people maintain so many varieties on the floodplain is unclear, but it points to the largely unrecognized potential of crop gene banks that peasant farmers have nurtured. Hundreds of years of culling and selection underlie the great variety of flavors, textures, colors, and growth habits of manioc now cultivated on the floodplain.

Both sweet and bitter varieties of manioc are processed into flour on the floodplain. To make flour, the tubers (particularly of the bitter manioc, which is toxic if unprocessed) are soaked for a few days in water and then pounded in a wooden trough. The cream-colored dough is then squeezed in a press, usually the traditional sleevelike press called tipiti. The squeezed dough is next passed through a sieve to remove fibrous lumps. Each batch of partially dry dough is stirred on a hot griddle for several hours. The cooked and crunchy flour can be stored without refrigeration for months.

The juice squeezed from manioc dough is usually saved to make a special sauce. This bright yellow sauce, called tucupi, is relished in fish stews. The starchy component of the juice, tapioca, can be served as light, pellet-sized balls or as a thick, gummy paste. Both forms are widely appreciated in several regional dishes. Tapioca pellets add texture to juice, while the clear paste is stirred into fish stews and tacacá, a spicy soup featuring dried shrimp. Tapioca is also served as chewy pancakes, called beiju, particularly in stalls in marketplaces.

At first glance, manioc production on the fertile soils of the Amazon may not appear a good use of the rich potential of the floodplain for food production. Higher value, and more nutritious cereals, would appear to be more appropriate for the floodplain. While there is certainly room for boosting cereal production along the Amazon, demand for manioc flour is also increasing. Manioc flour is a basic staple in both rural and urban areas, and with growing urban populations, market prospects for manioc products look bright.

LOOKING FOR FUTURE CROPS IN TODAY´S HOME GARDENS

Most floodplain inhabitants have home gardens hear their houses. The houses and home gardens themselves are often surrounded by fields of manioc, maize, and other crops. The main function of home gardens is to keep supplementary foods, medicinals, and flowers for decoration.

Farmers sometimes leave native trees when clearing forest or old second growth for their homes, if these trees are regarded as useful. These forest relics often produce seedlings in the newly open space. And these seedlings, in turn, may be tended in what will become home gardens. Another way that wild species enter the protodomestication stage is when valued seedlings sprout spontaneously in house yards, either as a result of natural dispersion or from inedible seeds discarded by family members while preparing or eating fleshy fruits at home.

Home gardens are thus propitious gene banks for promising new crops in Amazonia, and they can serve as sources of improved varieties for plantations and perennial cropping systems. Unfortunately, when attention turns to agricultural research and development in a region, home gardens are often deemed insignificant. Yet home gardens contain extraordinary diversity. They are often staging areas for trying out introduced crops, and they serve as arenas for domesticating wild plants from forest or successionary communities. By planting a few individuals of an unknown crop in the backyard, a farmer makes a minimal investment while observing its performance.

Home gardens are currently the most common and widespread form of tree farming on the Amazon floodplain. They provide tantalizing hints of what the landscape could look like, given modest investment. Besides home gardens, two other main tree-farming systems have evolved on the middle and lower Amazon. Over time, farmers have planted fruit trees in old cacao and rubber stands, whose value plummeted with the decline in the export market. Careiro Island near Manaus is marked by this enrichment. The other tree-farming system is for açaí. As described in the previous chapter, people have deliberately enriched the estuarine forests with açaí. But both forms of agroforestry together occupy just a tiny part of the Amazon flood-

A grove of bananas cultivated on a natural levee. *Bananas are a favorite crop on the higher levees bordering Amazon River channels or former channels. Although forests are razed to plant bananas, this form of farming is less destructive of the fisheries than is ranching because at least the floating meadows that border the levees remain intact.*

plain. Cacao and rubber stands enriched with fruit trees are increasingly rare, as they are felled to make way for cattle ranches and other land uses.

Large areas of underutilized floodplain, particularly areas degraded by livestock, could be rehabilitated with orchards planted with a mix of economically valuable species. A clue to how this might be accomplished can be gained by examining the role of home gardens in domesticating new species that provide food for people or bait for fish.

Home gardens contain a mixture of trees, bushes, and some annual plants, such as ornamentals and medicinals. One might not expect much diversity of woody crops in an environment that is mostly flooded for several months each year, yet home gardens on the floodplain are surprisingly rich in species. An inventory of perennials in just ten home gardens on the Amazon floodplain near Manaus and Itacoatiara revealed sixty-one woody species. Some home gardens on the Amazon floodplain contain twenty different shrubs and trees in addition to a variety of herbs and flowers.

The area occupied by a home garden is generally small, usually less than a hectare. And a home garden often grades into adjacent fields of maize or vegetables. Even caretakers on ranches will often maintain a home garden with medicinal plants in raised platforms and a surrounding orchard for shade, fruits, nuts, and other products. Several plants in home gardens, such as the cannonball tree, provide food for pigs, chickens, and ducks. Mango, banana, papaya, coconut, guava, genipap, and cupuaçu are especially common in home gardens. In addition to providing nutritional supplements, building materials, and medicines, home gardens also generate cash income. Bananas, in particular, are a significant cash crop along the floodplain.

With regard to plant domestication, home gardens display some interesting traits. First, a good part of the home garden owes more to selective weeding than to selective planting. Farmers simply protect whatever of value happens to sprout spontaneously. The tending of spontaneous plants is sometimes the first step toward their domestication. Trees that are welcome may be spared when the home is established. Other individuals or species are spared when they fortuitously invade the now-open landscape.

Trees do not always have to provide something edible in order to be spared. Trees that provide fruits for fish baits, such as the uruazeiro, are often left. Such practices could have much larger implications, as the quest for finding fruits that could be cultivated as food for fish farms is now getting under way. Second, some wild plants are right now in the early stages of domestication in the home gardens of Amazonia. The cannonball tree is such an example because of its value for livestock feed. Home gardens are thus likely to prove fertile hunting grounds for new crops for a wide variety of purposes.

UNCOVERING THE TREASURE

Most of the original floodplain rainforest along the middle and lower Amazon River has been destroyed. The giant trees have been felled; the floating meadows are being trampled by livestock. Today only remnants of the original floodplain ecosystems remain along more than two thousand kilometers of river. This devastation begins well upstream of Manaus (beyond the mouth of the Rio Purus) and continues all the way to the mouth of the Rio Xingu. Downstream of the Rio Xingu the tidal flooded forests have not yet been as hard hit, but their future is insecure. Large-scale logging of these estuarine forests has begun.

Although quantitative studies have not yet been made of the effects of floodplain deforestation, it is already clear that biodiversity is being destroyed on a large scale. Hundreds of plant species are now missing from the formerly rich floodplain areas that suffered deforestation. The animal species that have vanished from the floodplain of the middle and lower Amazon River probably number in the thousands if invertebrates are considered. How much biodiversity still exists in the remaining patches of forest is unknown. But it is well recognized that excessive fragmentation can deplete biodiversity even in untouched areas. If the present trend of deforestation continues, nearly all of the floodplain rainforest along the middle and lower Amazon River will be gone within just a few decades.

For most major plant and animal groups, significant differences in species composition grade along an axis defined by the main river. Many species found in the eastern floodplain are not present in the west, and vice versa. And there are great regional differences in the abundance of individual species that do span the full reach. For example, the Amazon region between Santarém and the Atlantic once supported enormous wading bird populations, but these are being greatly reduced because of nest habitat destruction and hunting pressure. None of the remaining nest sites is at present protected. Emilio Goeldi, a famous naturalist at the end of the last century, warned of the need to protect the bird life of the lower Amazon.

That the preservation of even relatively small floodplain areas is important to species protection is illustrated by the rich wildlife in the Anavilhanas Biological Reserve of the lower Rio Negro. Paulo Nogueira Neto, a former head of Brazil's environmental agency, played a powerful role in convincing government officials, the public, and conservationists from abroad of the need to incorporate biologically rich but small areas, in addition to large tracts of rainforest, into the reserve system. The fact that only small remnants of floodplain forest are now left along much of the middle and lower Amazon means that it is urgent to place at least some of these in reserves. No more than about five years remain for such action along much of the middle and lower Amazon. After that, even the remnants will be gone. The issue then will be restoration, not preservation.

A CALL FOR "FISH FORESTS" AND "FISH MEADOWS"

Present-day use of the Amazon River floodplain in many ways mirrors the pattern and consequences of settlement along the Transamazon Highway. In both cases massive deforestation has not led to a sustainable resource base or to an alleviation of poverty.

The main economic force driving the destruction of floodplain habitats is cattle and buffalo ranching. Timber extraction, cacao plantations, jute fields, and vegetable farming are minimal threats, compared to livestock operations. Large-scale livestock ranching has most affected the lower and middle Amazon, but it is also moving upstream. Water buffalo, which are better adapted to the floodplains than are cattle, are becoming more numerous (and destructive) each year.

Almost no scientific research, however, has been done on the effects of livestock ranching on the biodiversity and ecosystem health of Amazonian floodplains. Yet livestock ranching is the most dangerous land use that these ecosystems have ever faced.

The dependence of most commercial fish species on floodplain forests and floating meadows should stand as a compelling economic argument for the protection of these habitats. Luxuriant floating meadows are vital nurseries for the young of many commercial species. Later in life, these same fishes prowl the flooded forests for the prolific fruits, seeds, and insects that annually rain down upon them. Their diet not only produces flesh and fat during the floods; it also sustains them for months when the ebbing waters force their retreat into river channels.

Fish is still the cheapest and most important source of animal protein in the central Amazon Basin. There is no evidence that the wholesale conversion of the floodplain into livestock pasture will produce a more affordable source of animal protein than do the local fisheries. In fact, the dismantling of floodplain habitats will probably destroy many of the commercial fisheries. Needless to say, the loss of flooded forest also eliminates most of the biological richness native to this region, which includes a wealth of

The primary challenge to floodplain conservation.
Ranching of cattle and water buffalo poses the biggest threat to floodplain fisheries and overall biodiversity.

plant species, mammals, birds, amphibians, reptiles, and uncountable invertebrates. Many of these species are found nowhere else.

It would be absurd to argue that the Amazon River floodplain should not be developed in any way and that there is absolutely no role for livestock. Livestock on the floodplain are here to stay. But there is a big ecological difference between a local farmer or fisherman who chooses to run a few head of cattle and a rancher who makes cattle his business.

Every effort should be taken to prevent the wholesale conversion of flooded forests into pastures. Furthermore, floating meadows also need protection, as they are important habitats for fish. Large tracts of what we like to call *fish forests* and *fish meadows* should be set aside. Their designation as such would make it clear that the conservation gain would unquestionably accrue to local and regional economies. One need not be possessed of biophilia or concerned about species extinctions to support establishment of fish forests and fish meadows. But a spin-off benefit would be the protection of other biodiversity as well.

To this end, forest remnants still large enough to protect the fisheries in the long run need to be identified immediately as a first step toward selecting areas for protection and management. The most critical region for conducting surveys of potential fish forests and fish meadows centers on the middle and lower Amazon because of the massive deforestation that has already taken place there.

At present there are no floodplain parks or reserves in the lower 2,500 kilometers of the Amazon River. Moving upstream, the first reserve of any kind is the important Mamirauá Ecological Reserve. This reserve of 11,000 square kilometers was designated in 1990. It embraces a large delta-shaped area bounded on one side by the Amazon River (called Rio Solimões in this region) and by the Rio Japurá on the other. At least a third of the reserve is flooded forest. A research team headed by Márcio Ayres of the Goeldi Museum and of the Wildlife Conservation Society of the New York Zoological Society is now carrying out detailed biological and anthropological studies in the reserve.

The Mamirauá Ecological Reserve is important not only for its size and location; it is also a quest to make the ideals of preservation that appear on paper effective on the ground. Consider: in neither Brazil nor Peru has human settlement been both prohibited and enforced in any parks or reserves on the Amazon River floodplain. The philosophical thrust of the Mamirauá project, rather, is to integrate the local peoples (numbering about 4,500) into the reserve's management. They will not be removed. The Mamirauá project is thus a laudable test case to see if local peoples, without the threat of state or federal intervention, can be educated and convinced to abide by restrictions deemed necessary to protect biodiversity while at the same time continuing their use of it.

In our view, the next step for conservation would be to specifically tie reserves to economic ends. Protected areas would more than just accommodate residents; they would be tools for sustaining local and regional economies. Because the fisheries offer the best prospect for sustainable and significant economic returns, designated fish forests and fish meadows would be the way to proceed.

THE CHALLENGING TASK OF MANAGING THE FISHERIES

Of all the aquatic and floodplain resources harvested from the wild in the Amazon Basin, fish is unquestionably the most valuable in economic terms. This statement holds, whether one is assessing the interests of rural peasants, small communities, urban centers, Amazonian states, or the federal government.

Even floodplain farmers and many of the peasants that primarily tend livestock depend on fisheries for food and extra income. The importance of the fish resource to both urban and rural peoples has not, however, been satisfactorily addressed in government policies. One great impediment to government initiative is the matter of exactly who owns the fish resource.

Technically, federal governments own the fish resource, no matter where it is located. Many rural communities nevertheless believe that they themselves own the fish resource, and that they have the right to decide and profit from the sale of fishing rights to commercial interests. In contrast, commercial fishermen often feel it is their right to exploit any waters that become connected to the main river during the floods. Before implementing any broad-based management model, it is thus imperative to consider the many different circumstances under which de facto and de jure land and water ownership or occupation have evolved, and their effects on the fisheries of Amazonian floodplains. Many factors are important—such as distance from large urban centers, size of the local fishing fleet, large versus small de facto landholdings, and human population densities.

Unfortunately, current management of Amazonian fisheries will likely consign the fish resource to the boom-and-bust style of economy that this region has suffered since the very first days of the jute business. After thousands of years of well serving local needs, Amazonian fish resources in the 1960s bore the added burdens of provisioning growing cities with animal protein and of meeting export demand. The boom was on. By the early 1990s, however, the bust became imminent—at least for the preferred and first-class stocks, though fish continues to be the cheapest source of animal protein for most of the region.

During the 1960s and much of the 1970s, commercial fishermen, and especially those of the large Manaus fleet, operated in waters that were then largely uncontrolled either by government officials or by local residents. The Manaus fleet expanded to encompass most of the large state of Amazonas. Market forces alone dictated the development of these fisheries, and little thought was invested in managing the resource. The industrial fleet operating in the open waters of the estuary has never been regulated in any meaningful way. Now that preferred fish stocks are becoming rarer,

Forests and the fisheries. *Healthy floodplain forests are essential for sustainable fisheries throughout the Amazon.*

smaller towns and local communities, both along the inland rivers and in the estuary, have begun to prohibit, or at least attempt to control, outside fishermen.

The human population in the Amazon, and especially along the rivers, is largely urban. It is thus in cities and towns that most of the fish catch is consumed. Rural communities, however, have begun to demand greater control over local floodplain lakes and certain stretches of river. Local assertiveness is not surprising, in view of the fact that the federal and state governments have failed to take the lead in implementing management programs for the fisheries.

The principal regulatory measures attempted by the federal, state, and local governments in the Brazilian Amazon to control inland fishing have been outright closure during the high-water spawning season and size limits on individual species. The floods naturally and on their own suppress fishing effort because most species disperse widely in floodplain waters. Some of the migratory species, however, can be heavily exploited during the high-water period. Gillnet technology has made it possible for commercial fishermen to operate economically even during times when fish are widely dispersed.

Overall, the high-water prohibition has been largely ignored. Also, it is unclear from an ecological view whether this measure alone would protect several of the most important species from overexploitation. If fishing effort is excessive during the season of low water, the end result will be overexploitation of some stocks. A visit to almost any Amazonian fish market reveals that the second measure attempted—size minimums—is neither obeyed nor enforced. In fact, juvenile fish, many of which are technically illegal to market, are among the most important size classes sold in many towns and cities along the Amazon River.

The best management plan for any fisheries anywhere in the world would include the largest geographical area possible. This is especially true in the Amazon Basin, where migratory species account for the bulk of the commercial catch. The most extreme example of this is the fisheries for many species of large catfish that migrate up the Amazon River from the estuary. Migratory catfish are first heavily exploited by industrial operations in the estuary and then by as many as six different urban fishing fleets as they make their way up the Amazon River. Scientists from the Goeldi Museum in Belém and the Rainforest Alliance are now developing computer models to ascertain how catfish populations may be affected under a variety of harvesting scenarios up and down the long river system. This work will be vital for best balancing today's economic pressures against longer-term sustainability of the resource on which all interests ultimately depend.

Another important example of the huge areas embraced by some species is the case of the tambaqui, the most important food fish in the central Amazon. This fruit-eating characin can grow up to a meter long. The largest tambaqui are found upstream. Smaller and smaller individuals dominate along a gradient moving downstream toward the lower Amazon. Overfishing or habitat destruction in one area will weaken populations in the other, no matter how well managed the latter may be. Any model for the management of migratory species must thus include their spawning

Fish, the most important source of animal protein in the floodplain.
Steps must be taken to protect floodplain forests as part of fisheries management programs. The fishes of the Amazon have evolved with flooded forests, and the law should evolve to recognize this.

areas, nursery grounds, feeding habitats of adult fish, and seasonal movements related to river-level fluctuation.

The massive deforestation of Amazonian floodplains during the last three decades, and especially along the main river, has made it clear that very large areas must be considered when attempting to assess fish habitat destruction. We still do not know the minimum areas of flooded forests and floating meadows (the two most complex types of habitat) that fish populations need for breeding, nurseries, feeding, and protection from predators, including fishermen. It is unlikely that minimum habitat requirements can be determined from local studies alone. Large floodplain areas need to be surveyed in order to assess damage and to determine where important habitats still remain. Likewise, present land uses on the floodplains should be evaluated, since fisheries habitat protection must be weighed against these alternatives. Overall, the tendency for researchers and managers to focus on relatively small areas must be balanced by broader-scale perspectives. Both will prove vital for shaping effective management programs for fisheries and other resources.

Overharvesting is not, however, the only threat to Amazonian fisheries. Exploitation of other resources can have indirect effects on fisheries. Hydroelectric development is one concern, but existing impoundments do not seem to have measurably harmed the commercial fisheries in general (although biodiversity of noncommercial species is another matter). This is because the most important commercial fisheries are centered on the Amazon River, its sediment-rich tributaries, and just the lower two hundred kilometers of the blackwater and clearwater tributaries. Four of the five dams built thus far have been on relatively small blackwater or clearwater rivers. The Rio Tocantins, which bears the largest dam (Tucuruí), is not a tributary of the Amazon (it empties directly into the estuary). Thirty more dams are in the planning stage, however. Some of these might seriously affect local fisheries.

Mercury pollution is now a serious concern in the Amazon. Only in the Rio Madeira is there evidence that mercury has yet entered the food chain of any of the main commercial fish species. Because of the seriousness of mercury contamination, however, studies are now under way to determine to what extent fisheries and other resources may have been contaminated in other areas as well. If alarming levels of mercury were found in any of the commercial fish species harvested along the middle reaches of the Amazon River itself, the fisheries industry would almost certainly decline.

Finally, the petroleum industry must be watched and regulated out of concern for the fisheries. The western Amazon Basin and estuary are now being actively explored and drilled for oil. Tankers transport oil up and down the Amazon River, and there have been spills in headwater areas as far upriver as Ecuador. There is no functional plan in place to deal with a serious oil spill.

THE TANGLED NET OF FLOODPLAIN LAND OWNERSHIP

An issue that will need to be tackled if conservation and economic development are to be reconciled on the floodplain is the confused status of land tenure. The Amazon floodplain is not commonly the scene of violent clashes over land rights. Most migrants to Amazonia are settling in southern Pará, Rondônia, Acre, and Mato Grosso. Also, floodplain investors are buying out small farmers and ranches, rather than acquiring dubious land titles and then expelling the occupants. Most small farmers, in fact, do not have land titles, but they are nevertheless bought out to avoid conflict.

In some cases, farmland on the floodplain is individually worked, but the occupied plots are claimed by a local community. The community in turn is overseen by a president and village committee. Some floodplain residents also farm upland plots, and these tend to be individually owned, with clear title to the land. With increasing interest in developing the floodplain, the vague nature of land titles may lead to problems in the future. This factor could serve as a disincentive for small farmers to invest in land improvements.

The physical demarcation of land on the floodplain is, however, a challenging task. The restless floodplain is ever changing its configuration. Even new maps can quickly become misleading. Islands disappear, while others are born. Now that cattle and water buffalo roam freely, the ownership issue is becoming even more confused.

To succeed, a management plan for Amazonian wetlands must include realistic and clear-cut policies defining who legally controls the rivers and floodplains. Technically, the Brazilian, Peruvian, Colombian, and Bolivian federal governments own

Confusions over land title. *The Amazon River floodplain has not been surveyed or zoned to indicate ownership and control. Most inhabitants have informal documents that they use as de facto land titles. Until surveys are made it will be very difficult to protect biodiversity.*

Amazonian rivers and floodplain waters. This includes the forests and wildlife found in, on, and along them as well. Responsibility for these federally owned resources has historically been divided among various government agencies. This is especially true in Brazil.

Until recently the Brazilian Institute of Forestry Development (IBDF) was vested with control over all forests and wildlife—*but not fishes.* The exclusion of fisheries from forest management was largely a vestige of North American and European models. But these models have their faults. For example, the United States has historically managed (and extensively logged) its northwestern forests without adequate provision for protecting the waterways into which salmon migrate and spawn. In Brazil the now-extinct Superintendency of Fisheries Development (SUDEPE) once had control over fisheries, although evidently no agency was effectively in charge of fish resources in general. The Maritime Ministry was charged with controlling open waters and registering all craft, including fishing boats.

In 1988 all resource management was transferred to the specially established Brazilian Institute of the Environment (IBAMA), in what was to be a strong and conservation-active federal government. Since then, owing to a weakened federal government, lack of funds, and insufficient trained personnel, IBAMA has been unable to carry out its charter. There is now much discussion of transferring considerable environmental authority to state and municipal governments. The ecological implications of decentralization have not, however, been adequately examined. Ecological consequences would undoubtedly vary in different regions because of the dominating effect of local politics in a landscape as vast as the Amazon.

Amazon floodplain ownership and resource management pose special problems because of seasonal or daily flooding. These "lands" can be both aquatic and terrestrial during the course of a year. Inhabitants often claim to "own" a patch of floodplain for half the year when it is mostly dry. When the floods return, these owners supposedly relinquish their hold, and the floodplain is then open to all, including commercial fishermen. "It's mine when dry, ours when wet," seems to be the refrain.

The main problem with this half-and-half approach is that the private "owner" of the floodplain has no incentive to curtail activities during the dry season that might be extremely detrimental to economic use during the season of public ownership. Notably, deforestation largely takes place during the low-water period when most of the floodplain is dry and thus "owned" by particular occupants. When the inundated floodplain reverts to de facto government ownership during the flood season, the very rainforest and other habitats on which fish and other wildlife once depended might be gone. Until the land ownership problem is settled, the Amazon floodplain will remain in grave danger.

DOUBTS ABOUT COMMUNITY-BASED MANAGEMENT

At present there are no federal parks or reserves along the Amazon River in Brazil. In fact, the only protected area is the Mamirauá Ecological Reserve in the western Amazon, and it is presided over by the state of Amazonas. International organizations (including the British Overseas Development Administration, the World Wildlife Fund, and the Wildlife Conservation Society) are financing the Mamirauá project. This conservation effort was undertaken with a commitment to intensively study the biodiversity, ecology, and anthropology of the designated area in order for local management programs to have a realistic biological and cultural foundation. The Mamirauá inventory will thus be the first large-scale survey of plants, animals, and sociological aspects done on any relatively natural floodplain area of the Amazon River.

One of the main working hypotheses of the Mamirauá project is that the local rural communities can be convinced to manage natural resources without strong federal or state policing actions. Project success depends on this assumption carrying through. Project supporters thus hope to demonstrate that local communities can indeed be integrated into an ecological reserve. Nevertheless, if the opposite turns out to be the case, that finding will itself be a significant conclusion for future planning of protected areas.

The Iara project, near Santarém, is also a community-based experiment. Unlike Mamirauá, however, no large remnants of floodplain forest remain in the Santarém region, and biodiversity is greatly reduced. As it would be a daunting task to develop a model of what floodplain biodiversity in the Santarém area was like before deforestation took place, the Iara project is instead concentrating mainly on the fisheries resource. Here the assumption is that the local communities can learn enough about fish ecology and production in order to establish meaningful exploitation limits for urban and rural fishermen. Furthermore, as at Mamirauá, it is assumed that the local communities can be convinced by researchers, rather than forced by policing authorities, to manage the resources.

Two factors may, however, undermine the best intentions of the founders and implementors of the Iara project. First, management is focused on regulation of fish takes, not on controlling land uses in the habitats on which fish ultimately depend. Without a strong habitat component to the management plan for the highly disturbed parts of the Amazon floodplain—which would largely mean reforestation programs and protection of floating meadows—local communities may not have a fair chance to protect and sustainably manage the fisheries in their areas. A second factor undermining success is that many of the species are migratory and thus depend on the habitat health and harvest restrictions in areas far beyond the control of communities enrolled in the Iara project. If communities are to successfully manage migratory fisheries, they may have to form inter-regional networks. And they will surely need to rely on data supplied by scientists about migration, productivity, and other aspects of ecology. Then most of the communities would have to jointly set catch limits and perhaps initiate habitat protection schemes.

Overall, because the fisheries are so interconnected in the Amazon, we are doubtful that community-based approaches to resource management will be able to succeed on their own—that is, without some type of outside enforcement.

RECONCILING CONSERVATION AND DEVELOPMENT

The Amazon is too large, and financial and human scientific resources have been too scarce, for any institution yet to design adequate programs reconciling conservation and development anywhere in the Amazon—particularly in the gravely threatened floodplains. Any plan to conserve biodiversity and to sustainably manage natural resources must take into account the history, needs, and incentives that transformed the floodplain into what it is today and the forces that will drive future changes. A thorough understanding of how people now use and abuse biological resources and habitats is an essential prelude to any interventions—whether to protect forests, to induce greater use of native plants and animals in farming (and turn away from exotics), or to increase agricultural yields, whatever the source.

At present very little is known about the diversity of land uses that have evolved along the Amazon River. Some of these, such as vegetable growing, have actually led to more intensive but less extensive habitat alterations—benefiting both conservation and development. Yet the marked increase of cattle and water buffalo ranching in the floodplain could overwhelm these gains before they can be studied and improved for wider use.

After two decades of focus on the uplands, policy makers and investors are looking increasingly to the Amazon floodplain for enhanced development. The traditional development bias against the floodplain—owing to its cycling between land and water—has until recently checked some of the large-scale and high-impact forms of resource exploitation. But that is now changing.

For decades, a few local politicians and scientists have recognized that the Amazon floodplain is especially propitious for agricultural development because of its rich, annually replenished soils and ease of access. Anthropologists, such as Betty Meggars of the Smithsonian Institution, concluded that the floodplains would, in fact, be better for agricultural development than would the uplands. Any attempt to turn the Amazon floodplain into a giant pasture for exotic species should weigh the injury to biodiversity against the real benefits (and costs) likely to be felt by the local peoples—including those who depend on the fisheries. To date there is no evidence that livestock ranching or any other large-scale land use that extensively alters the floodplain forests and floating meadows is significantly raising the standard of living of most of the people living in these areas. Yet large-scale degradation of the environment continues, much as it did along the Transamazon Highway.

What few scientists and politicians have confronted is that Amazonian floodplains have their own unique complement of both plants and animals. Although floodplain habitats constitute just 2 to 3 percent of the Amazon Basin, they are on the exposed edge of the rainforest. Access to these habitats from the rivers has been easy, and human activities along the main Amazon have led to serious fragmentation of floodplain forests. Recently, large-scale introduction of livestock has also provoked the destruction of floating meadows. Yet these forest and meadow habitats, and the rich fish life associated with them, hold one of the best bets for economic development in the Amazon. Here lie the treasures delivered by the floods of fortune.

We believe that one-dimensional approaches to conservation and development of the Amazon floodplain should be avoided. For example, placing too much faith in the community management approach to resource conservation is unwise. If rural communities fail to manage resources, they alone might be blamed for not protecting the environment, when in fact this is also one of the duties of modern governments. Federal and state governments should support community experiments in resource management, but at the same time every attempt should be made to establish parks and reserves where deforestation, logging, farming, and livestock ranching are not allowed. There is simply too little evidence to support the idea that community management alone will provide the long-term means to guarantee the future of floodplain resources and wildlife.

A multifaceted and more dynamic approach to Amazon floodplain management should involve a variety of stakeholders, including urban centers, rural communities,

Ominous shadows.
Burned tree stumps cast menacing shadows on the floodplains of the Amazon River.

What fate for flooded forests?
The future of the flooded forests depends on finding ways to reconcile development with conservation.

commercial interests, conservation organizations, and academic institutions. One of the biggest challenges to conservation efforts is to draw in these various interests for constructive solutions. We believe that the most powerful "hook" to begin this process is with the fish resource, because of its overwhelming economic and nutritional importance and its potential. The health of the fish resource provides a measure against which farming, cattle ranching, logging, and other floodplain activities can be judged. If too great an environmental toll is taken on the floodplain forests and floating meadows, then almost everyone will be losers and there can be little hope of economic gain in the long run.

Our purpose has been to focus attention sharply on the Amazon floodplain—for, in doing so, steps might be taken to give this amazing environment its historical, present, and future due, and to bring about a better life for the people who depend on it.

COMMON AND SCIENTIFIC NAMES OF PLANTS AND ANIMALS

ENGLISH	BRAZILIAN	SCIENTIFIC
PLANTS		
American oil palm	caiaué	*Elaeis oleifera*
andiroba	andiroba	*Carapa guianensis*
arum	aninga	*Montrichardia arborescens*
assacu	assacu	*Hura crepitans*
assaí palm	açaí	*Euterpe oleracea*
bacuri	bacuri	*Platonia insignis*
cacao	cacauo	*Theobroma cacao*
camapu	camapu	*Physalis angulata*
cannonball tree	castanha de macaco	*Couroupita guianensis*
cedar	cedro	*Cedrela odorata*
cocoyam	cocoyam	*Xanthosoma* sp.
copaíba	copaíba	*Copaifera mulitjuga*
cupuaçu	cupuaçu	*Theobroma grandiflorum*
false cinnamon	canela	*Ocotea quixos*
false clove	cravo	*Dicypellium caryophyllatum*
genipap	jenipapo	*Genipa americana*
giant water lily	Vitória régia	*Victoria amazonica*
guarumã	guarumã	*Ischnosiphon obliquus*
hogplum, yellow mombim	taperebá, cajá	*Spondias mombim*
ingá	ingá	*Inga* spp.
itaúba	itaúba	*Mezilaurus itauba*
jacareúba	jacareúba	*Calophyllum brasiliensis*
jacitara	jacitara	*Desmoncus polyacanthos*
jauari	jauari	*Astroycaryum jauari*

ENGLISH	BRAZILIAN	SCIENTIFIC
PLANTS *continued*		
junco	junco	*Cyperus* sp.
jupati	jupati	*Raphia vinifera*
kapok	sumaúma	*Ceiba pentranda*
mahogany	mogno	*Swietenia macrophylla*
mamorana	mamorana	*Catostema albuquerquei*
manioc	mandioca	*Manihot esculenta*
marajá	marajá	*Bactris* sp.
mauritia	buriti, miriti	*Mauritia flexuosa*
mulato-wood	pau mulato	*Peltogyne paniculata*
mumbaca	mumbaca	*Astrocaryum* sp.
munguba, silk-cotton tree	munguba	*Pseudobombax munguba*
muri	muri	*Paspalum fasciatum*
murumuru	murumuru	*Astrocaryum murumuru*
pemembeca	pemembeca	*Paspalum repens*
piranha-tree	piranheira	*Piranhea trifoliata*
root beer vine	sarsaparilla	*Smilax papyracea*
rubber	seringa	*Hevea brasiliensis*
sapucaia	sapucaia	*Lecythis pisonis*
socoró	socoró	*Mouriria* cf. *ulei*
soursop	graviola	*Annona muricata*
tucumã	tucumã	*Astrocaryum aculeatum*, *A. vulgare*
uruazeiro	uruazeiro	*Cordia* sp.
urucuri	urucuri	*Attalea phalerata*
uxi	uxi	*Endopleura uchi*
vanilla	vanilão	*Vanilla* sp.
virola	ucuúba	*Virola surinamensis*
wild rice	arroz	*Orzya* spp.
willow	oeirana	*Salix martiana*
BIRDS		
ani	anu	*Crotophaga* spp.
cacique	japim	*Cacicus cela*
cattle egret	garça vaqueira	*Bubulcus ibis*
harpy eagle	gavião real	*Harpia harpyja*
herons	socó	Family Ardeidae
hoatzin	cigana	*Opisthocomus hoazin*
ibises	curicaca, guará	Family Threskiornithidae
limpkin	carão	*Aramus guarauna*
macaws	araras	*Ara* spp.
maguari stork	maguari	*Ciconia maguari*
muscovy duck	pato-do-mato	*Cairina moschata*
parakeets	periquitos	*Brotogris* spp.
parrotlets	periquitos	*Forpus* spp.
parrots	papagaios	*Amazona* spp.

ENGLISH	BRAZILIAN	SCIENTIFIC
BIRDS *continued*		
toucan	tucano	Family Ramphastidae
troupial	corrupião	*Icterus icterus*
tyrant flycatchers	bem-ti-vi	Family Tyrannidae
Amazonian umbrellabird	anambé	*Cephalopterus ornatus*
MAMMALS		
bôto dolphin	bôto	*Inia geoffrensis*
capuchin monkey	macaco prego	*Cebus* spp.
capybara	capivara	*Hydrochaeris hydrochaeris*
howler monkey	guariba	*Alouatta* spp.
manatee	peixe-boi	*Trichechus inunguis*
peccary	catitu, porco da mata	*Tayassu* spp.
pygmy marmoset	sagüi	*Cebuella pygmaea*
squirrel monkey	macaco de cheiro	*Saimiri sciureus*
tapir	anta	*Tapirus terrestris*
titi monkey	zogue-zogue	*Callicebus* spp.
tucuxi dolphin	tucuxi	*Sotalia fluviatilis*
uacari monkey	uacari	*Cacajao* spp.
FISH		
characins	piabas	Order Characiformes
cichlids	cichlídeos	Family Cichlidae
curimatá	curimatã	*Prochilodus nigricans*
doradid catfishes	bacu	Family Doradidae
dourada catfish	dourada	*Brachyplatystoma flavicans*
drums	pescadas	Family Sciaenidae
herring	apapá	*Pellona* spp.
jaraqui	jaraqui	*Semaprochilodus* spp.
mapará catfish	mapará	*Hypophthalmus* spp.
pimelodid catfishes	bagres	Family Pimelodidae
piramutaba catfish	piramutaba	*Brachyplatysoma vaillantii*
piranha	piranha	Subfamily Serrasalminae
pirapitinga	piripitinga	*Piaractus nigripinnus*
pirarara catfish	pirarara	*Phractocephalus hemiliopterus*
pirarucu, red-fish	pirarucu	*Arapaima gigas*
rock-bacu	bacu pedra	*Lithodoras dorsalis*
tambaqui, black pacu	tambaqui	*Colossoma macropomum*
tucunaré, peacock bass	tucunaré	*Cichla* spp.
REPTILES		
black caiman	jacaré-açu	*Melanosuchus niger*
colubrid (common) snakes	cobra comum	Family Colubridae
coral snake	cobra coral	Family Elapidae
giant river turtle	tartaruga	*Podocnemis expansa*
pit vipers	jararaca, surucucu	Family Viperidae

ENGLISH	BRAZILIAN	SCIENTIFIC
REPTILES *continued*		
spectacled caiman	jacaretinga	*Caiman crocodilus*
teiid lizard	jacuruxi	Family Teiidae
AMPHIBIANS		
Bufo toad	sapo	*Bufo* spp.
Pipa frog	pipa	*Pipa* spp.

BIBLIOGRAPHY

Adis, J. 1983. Comparative ecological studies of the terrestrial arthropod fauna in central Amazonian inundation-forests. *Amazoniana* 7 (2): 87–173.

Agassiz, L. and E. Agassiz. 1867. *A Journey in Brazil.* Boston: Ticknor & Fields.

Alden, D. 1976. The significance of cacao production in the Amazon region during the late colonial period: An essay in comparative economic history. *Proceedings of the American Philosophical Society* 120 (2): 116.

Anderson, A. B. 1988. Use and management of native forests dominated by açaí palm (*Euterpe oleracea* Mart.) in the Amazon estuary. *Advances in Economic Botany* 6:144–154.

———. 1994. *Impactos Ecológicos e Sócio-Econômicos da Exploração Selectiva de Virola no Estuário Amazônico.* Washington, D.C.: World Wildlife Fund.

Anderson, A. B. and E. M. Ioris. 1992. Valuing the rain forest: Economic strategies by small-scale forest extractivists in the Amazon estuary. *Human Ecology* 20 (3): 337–369.

Anderson, R. L. 1976. Following curupira: Colonization and migration in Pará, 1758 to 1930, as a study of settlement of the humid tropics. Ph.D. diss. University of California, Davis.

Araújo-Lima, C. A. R. M., B. R. Forsberg, and R. Victoria. 1986. Energy sources for detritivorous fishes in the Amazon. *Science* 234:1256–1258.

Ayres, J. M. 1993. *As Matas de Várzea do Mamirauá.* Brasília: CNPq/Sociedade Civil de Mamirauá.

———. 1986. Uakaris and Amazonian flooded forest. Ph.D. diss. Cambridge University, England.

Bahri, S. 1992. L'agroforesterie—une alternative pour le développement de la plaine alluviale de l'Amazonie: L'exemple de l'île de Careiro. Ph.D. diss. Université de Montpellier, France.

Balick, M. J., ed. 1988. The palm—tree of life: Biology, utilization, and conservation. *Advances in Economic Botany.* New York: New York Botanical Garden.

Balick, M. J. and H. T. Beck. 1990. *Useful Palms of the World: A Synoptic Bibliography.* New York: Columbia University Press.

Barthem, R. B. 1985. Ocorrência, distribuição e biologia dos peixes da Baía de Marajó, Estuário Amazônico. *Boletim do Museu Paraense Emilio Goeldi, Zoologia* 15 (12): 49–69.

———. 1990. Descricão da pesca da piramutaba. *Boletim do Museu Paraense Emilio Goeldi, Antropologia* 6:117–130.

Barthem, R. B., M. C. L. B. Ribeiro, and M. Petrere. 1991. Life strategies of some long distance migratory catfish in relation to hydroelectric dams in the Amazon Basin. *Biological Conservation* 55:339–345.

Bates, H. W. 1863. *The Naturalist on the River Amazons.* London: John Murray.

Bayley, P. B. 1982. Central Amazon fish populations: Biomass, production, and some dynamic characteristics. Ph.D. diss. Dalhousie University, Canada.

———. 1986. Aquatic productivity in the central Amazon "várzea" in the context of the fishery yield. *Proceedings of the First Symposium on the Humid Tropics.* Belém: Embrapa, pp. 325–334.

———. 1988. Factors affecting growth rates of young tropical floodplain fishes: Seasonality and density-dependence. *Environmental Biology of Fishes* 21 (2): 127–142.

Bayley, P. B. and M. Petrere. 1989. Amazon fisheries: Assessment methods, current status, and management options. *Canadian Special Publications: Fisheries and Aquatic Science* 106:385–398.

Biery-Hamilton, G. M. 1987. Coping with change: The impact of the Tucuruí dam on an Amazonian community. Master's thesis. University of Florida, Gainesville.

Bigarella, J. J. and A. M. M. Ferreira. 1985. Amazonian geology and the Pleistocene and the Cenozoic environments and paleoclimates. In G. T. Prance and T. E. Lovejoy, eds., *Key Environments: Amazonia,* pp. 49–71. Oxford: Pergamon Press.

Biller, D. 1994. Informal gold mining and mercury pollution in Brazil. *Policy Research Working Paper of the World Bank* (May 1994): 1–28.

Black, G. A., T. Dobzhansky, and C. Pavan. 1950. Some attempts to estimate species diversity and population density of trees in Amazonian forests. *Botanical Gazette* 111:413–425.

Bodmer, R. E. 1989. Frugivory in Amazonian Artiodactyla: Evidence for the evolution of the ruminant stomach. *Journal of the Zoological Society of London* 219:457–467.

———. 1991. Strategies of seed dispersal and seed predation in Amazonian ungulates. *Biotropica* 23 (3): 255–261.

Bohlke, J. E., S. H. Weitzman, and N. A. Menezes. 1978. Estado atual da sistemática dos peixes de água doce da América do Sul. *Acta Amazonica* 8 (4): 657–678.

Boischio, A. A. P. 1993. Exposicão ao mercúrio orgânico em populacões reiberinhas do Alto Madeira, Rondônia, 1991: Resultados preliminares. *Cadernos de Saúde Pública* 9 (2): 1–6.

Boxer, C. R. 1975. *The Golden Age of Brazil.* Berkeley: University of California Press.

Burns, B. 1965. Manaus, 1910: Portrait of a boom town. *Journal of Inter-American Studies* 7 (3): 400–421.

Bush, M. A., D. R. Piperno, and P. A. Colinvaux. 1989. A 6,000 year history of Amazonian maize cultivation. *Nature* 340:303–305.

Campbell, D. G., D. C. Daly, and G. T. Prance. 1986. Quantitative ecological inventory of terra firme and várzea tropical forest on the Rio Xingu, Brazilian Amazon. *Brittonia* 38 (4): 369–393.

Campbell, D. G., J. L. Stone, and A. Rosas. 1992. A comparison of the phytosociology and dynamics of three floodplain (várzea) forests of known ages, Rio Juruá, western Brazilian Amazon. *Botanical Journal of the Linnean Society* 108:213–237.

Carvalho, J. L. and B. Merona. 1986. Estudos sobre dois peixes migratórios do baixo Tocantins, antes do fechamento da barragem de Tucuruí. *Amazoniana* 9:595–607.

Cleary, D. 1990. *Anatomy of the Amazon Gold Rush.* Iowa City: University of Iowa Press.

Coimbra-Filho, A. F. and R. A. Mittermeier, eds. 1981. *Ecology and Behavior of Neotropical Primates.* Rio de Janeiro: Academia Brasileira de Ciências.

Colinvaux, P. A. 1989. The past and future Amazon. *Scientific American* 260 (5): 102–108.

Cunha, O. R. and F. P. Nascimento. 1978. Ofídios da Amazônia. X. As cobras da região leste do Pará. *Publicacões Avulsas, Museu Paraense Emilio Goeldi* 31:1–218.

Dean, W. 1987. *Brazil and the Struggle for Rubber.* Cambridge, England: Cambridge University Press.

Dixon, J. R. 1979. Origin and distribution of reptiles in lowland tropical rainforests of South America. In W. E. Duellman, ed., *The South American Herpetofauna: Its Origin, Evolution, and Dispersal,* pp. 217–248. Lawrence, Kansas: University of Kansas Press.

Emmons, L. H. 1990. *Neotropical Rainforest Mammals: A Field Guide.* Chicago: University of Chicago Press.

Erwin, T. L. 1988. The tropical forest canopy: The heart of biotic diversity. In E. O. Wilson and F. M. Peter, eds., *Biodiversity*, pp. 123–129. Washington, D.C.: National Academy Press.

Erwin, T. L. and J. Adis. 1982. Amazonian inundation forests: Their role as short-term refuges and generators of species richness and taxon pulses. In G. T. Prance, ed., *Biological Diversification in the Tropics*, pp. 358–374. New York: Columbia University Press.

Fearnside, P. M. 1990. Balbina: Lições trágicas da Amazônia. *Ciência Hoje* 11 (64): 34–43.

———. 1990. Deforestation in Brazilian Amazonia. In G. M. Woodwell, ed., *The Earth in Transition: Patterns and Processes of Biotic Impoverishment*, pp. 211–239. Cambridge, England: Cambridge University Press.

Feijão, A. and J. A. Pinto. 1990. *Garimpeiros in South America: The Amazon Gold Rush*. Pará: USAGAL/BM&F.

Ferreira, E. F. G. 1984. A ictiofauna da represa hidrelétrica de Curuá-Una, Santarém, Pará: I. Lista e distribuição das espécies. *Amazoniana* 8:351–363.

Foster, R. 1990. The floristic composition of the Manu floodplain forest. In A. H. Gentry, ed., *Four Neotropical Rainforests*, pp. 99–111. New Haven, Conn.: Yale University Press.

Fritz, S. 1922. *Journal of the Travels and Labours of Father Samuel Fritz in the River of the Amazons Between 1686 and 1723*. London: Hakluyt Society.

Gentil, J. M. L. 1988. A juta na agricultura de várzea na área de Santarém—Médio Amazonas. *Boletim do Museu Paraense Emilio Goeldi* 4 (2): 118–199.

Gentry, A. H. and J. Lopez-Parodi. 1980. Deforestation and increased flooding of the upper Amazon. *Science* 210:1354–1356.

Gentry, A. H. and J. Terborgh. 1990. Composition and dynamics of the Cocha Cashu mature floodplain forest. In A. H. Gentry, ed., *Four Neotropical Rainforests*, pp. 542–564. New Haven, Conn.: Yale University Press.

Goulding, M. 1980. *The Fishes and the Forest: Explorations in Amazonian Natural History*. Los Angeles: University of California Press.

———. 1981. *Man and Fisheries on an Amazon Frontier*. The Hague: Dr. W. Junk Publishers.

———. 1983. Amazonian fisheries. In E. F. Moran, ed., *The Dilemma of Amazonian Development*, pp. 189–210. Boulder, Colo.: Westview Press.

———. 1989. *Amazon: The Flooded Forest*. London: The BBC.

Goulding, M. and M. L. Carvalho. 1982. Life history and management of the tambaqui (*Colossoma macropomum*, Characidae): An important Amazonian food fish. *Revista Brasileira de Zoologia* 1 (2): 107–133.

Goulding, M., M. L. Carvalho, and E. G. Ferreira. 1988. *Rio Negro: Rich Life in Poor Water*. The Hague: SPB Academic Publishing.

Gribel, R. 1990. The Balbina disaster: A need to ask why? *The Ecologist* 20:133–135.

Guidon, N. and G. Delibrias. 1986. Carbon-14 dates point to man in the Americas 32,000 years ago. *Nature* 321:769–771.

Hecht, S. B. and A. Cockburn. 1989. *The Fate of the Forest: Developers, Destroyers, and Defenders of the Amazon*. London: Verso.

Heiser, C. B. 1985. *Of Plants and People*. Norman, Okla.: University of Oklahoma Press.

Hemming, J. 1978. *Red Gold: The Conquest of the Brazilian Indians, 1500–1760*. Cambridge: Harvard University Press.

———. 1987. *Amazon Frontier: The Defeat of the Brazilian Indians*. London: Macmillan.

Herrera, L. F., I. Cavelier, and C. Rodriguez. 1992. The technical transformation of an agricultural system in the Colombian Amazon. *World Archaeology* 24 (1): 98–113.

Heyer, W. R. and R. W. McDiarmid. 1975. Tadpoles, predation, and pond habitats in the tropics. *Biotropica* 7 (2): 100–111.

Hiraoka, M. 1985. Changing floodplain livelihood patterns in the Peruvian Amazon. *Tsukuba Studies in Human Geography* 3:243–275.

———. 1985. Floodplain farming in the Peruvian Amazon. *Geographical Review of Japan* 58 (Ser. B, 1): 1–23.

———. 1985. Mestizo subsistence in riparian Amazonia. *National Geographic Research* 1 (2): 236–246.

———. 1986. Zonation of mestizo farming systems in northeast Peru. *National Geographic Research* 2 (3): 354–371.

———. 1989. Agricultural systems on the floodplains of the Peruvian Amazon. In J. O. Browder, ed., *Fragile Lands of Latin America: Strategies for Sustainable Development*, pp. 75–99. Boulder, Colo.: Westview Press.

———. 1993. Caboclo and ribereño resource management in Amazonia: A review. In K. H. Redford and C. Padoch, eds., *Conservation of Neotropical Forests: Working from Traditional Resource Use*, pp. 134–157. New York: Columbia University Press.

Hödl, W. 1977. Call differences and calling site segregation in Anuran species from central Amazonian floating meadows. *Oecologia* 28:351–363.

Horgan, J. 1992. Early arrivals: Scientists argue over how old the New World is. *Scientific American* 266 (2): 17–20.

Junk, W. J. 1970. Investigations on the ecology and production-biology of the "floating meadows" (Paspalo-Echinochloetum) on the middle Amazon. Part I: The floating vegetation and its ecology. *Amazoniana* 2 (4): 449–496.

———. 1984. Ecology of the várzea, floodplain of Amazonian white water rivers. In H. Sioli, ed., *The Amazon: Limnology and Landscape Ecology of a Mighty Tropical River and Its Basin*, pp. 215–243. The Hague: Dr. W. Junk Publishers.

———. 1989. The use of Amazonian floodplains under an ecological perspective. *Interciencia* 14:317–322.

Junk, W. J. and K. Furch. 1985. The physical and chemical properties of Amazonian waters and their relationships with the biota. In G. T. Prance and T. E. Lovejoy, eds., *Key Environments: Amazonia*, pp. 3–17. New York: Pergamon Press.

Junk, W. J. and C. Howard-Williams. 1984. Ecology of aquatic macrophytes in Amazonia. In H. Sioli, ed., *The Amazon: Limnology and Landscape Ecology of a Mighty Tropical River and Its Basin*, pp. 269–293. The Hague: Dr. W. Junk Publishers.

Junk, W. J. and J. A. S. N. Mello. 1987. Impactos ecológicos de represas hidroelétricas na bacia amazônica brasileira. In G. Kohlhepp and A. Schrader, eds., *Homem e Natureza na Amazônia*, pp. 367–385. Tübingen, Germany: A. Tübingen Geographisch.

Junk, W. J., B. A. Robertson, and A. J. Darwich. 1981. Investigações limnológicas e ictiológicas em Curuá-Una, a primeira represa hidrelétrica na Amazônia central. *Acta Amazonica* 11:689–716.

Keller, F. 1875. *The Amazon and Madeira Rivers: Sketches and Descriptions from the Note-Book of an Explorer*. Philadelphia: J. B. Lippincott.

Kroner, A. and P. W. Layer. 1992. Crust formation and plate motion in the early Archean. *Science* 256:1405–1411.

Le Cointe, P. 1947. *Árvores e Plantas Úteis (Indígenas e Aclimadas)*. São Paulo: Companhia Editora Nacional.

Lima, R. 1956. A agricultura nas várzeas do estuário do Amazonas. *Boletim Técnico do Instituto Agronômica do Norte* 33:1–164.

Lovejoy, T. E. 1975. Bird diversity and abundance in Amazon forest communities. *The Living Bird* 13:127–191.

Lowe-McConnell, R. H. 1984. The status of studies on South American freshwater food fishes. In T. M. Zaret, ed., *Evolutionary Ecology of Neotropical Freshwater Fishes*, pp. 139–156. The Hague: Dr. W. Junk Publishers.

———. 1987. *Ecological Studies in Tropical Fish Communities*. Cambridge, England: Cambridge University Press.

McGrath, D. G., F. Castro, and C. Futemma. 1993. Fisheries and the evolution of resource management on the lower Amazon floodplain. *Human Ecology* 21 (2): 167–195.

Mahar, D. J. 1976. Fiscal incentives for regional development: A case study of the western Amazon Basin. *Journal of Inter-American Studies and World Affairs* 18 (3): 357–379.

———. 1979. *Frontier Development Policy in Brazil: A Study of Amazonia.* New York: Praeger.

———. 1989. *Government Policies and Deforestation in Brazil's Amazon Region.* Washington, D.C.: World Bank.

Mallas, J. and N. Benedicto. 1986. Mercury and gold mining in the Brazilian Amazon. *Ambio* 15:248–249.

Malm, O., W. C. Pfeiffer, and C. M. M. Souza. 1990. Mercury pollution due to gold mining in the Madeira River Basin, Brazil. *Ambio* 19:11–15.

Marques, J. R. F., R. S. Chaves, and D. G. McGrath. 1992. Exploracão sustentada da agricultura e pecuária de várzea. In *Simdamazonia: Seminario Internacional Sobre meio Ambiente, Pobreza e Desenvolvimento da Amazônia*, pp. 269–274. Belém: SECTMA and Governo do Estado do Pará PRODEPA.

Martinelli, L. A., J. R. Ferreira, and B. R. Forsberg. 1988. Mercury contamination in the Amazon: A gold rush consequence. *Ambio* 17:252–254.

Medina, J. T. 1988. *The Discovery of the Amazon.* New York: Dover.

Meggers, B. J. 1971. *Man and Culture in a Counterfeit Paradise.* Chicago: Aldine.

Merona, B. 1987. Aspectos ecológicos da ictiofauna no baixo Tocantins. *Acta Amazonica* 16/17: 109–124.

Mesquita, M. G. C. 1969. Terra-água no complexo amazônico e a organização da economia regional. In *Problemática da Amazônia.* Rio de Janeiro: Casa do Estudante do Brasil.

Ministério de Minas e Energia. 1992. Estudo dos impactos ambientais na reserva garimpeira do Tapajós, Estado do Pará. *Série Tecnologia Mineral, DNPM* 2 (2): 1–228.

Mittermeier, R. A., A. B. Rylands, and A. F. Coimbra-Filho, eds. 1988. *Ecology and Behavior of Neotropical Primates.* Washington, D.C.: World Wildlife Fund.

Nascimento, C. and A. Homma. 1984. *Amazônia: Meio Ambiente e Tecnologia Agrícola.* Belém: EMBRAPA.

National Research Council. 1981. *The Water Buffalo: New Prospects for an Underutilized Animal.* Washington, D.C.: National Academy Press.

Nriagu, J. O., W. C. Pfeiffer, and O. Malm. 1992. Mercury pollution in Brazil. *Nature* 356:746.

Odinetz-Collart, O. 1991. Tucuruí dam and the populations of the prawn *Macrobrachium amazonicum* in the lower Tocantins (PA-Brazil): A four-year study. *Archives of Hydrobiology* 122:213–227.

ORSTOM (Institut Français de Recherches pour le Développement en Coopération). 1988. *Conditions Ecologiques et Economiques de la Production d'une Ile de Várzea: L'île du Careiro.* Paris: ORSTOM.

Padoch, C. 1988. Aguaje (*Mauritia flexuosa* L.f.) in the economy of Iquitos, Peru. The palm-tree of life: Biology, utilization, and conservation. *Advances in Economic Botany* 6:241–224.

———. 1988. People of the floodplain and forest. In J. S. Denslow and C. Padoch, eds., *People of the Tropical Rain Forest*, pp. 127–140. Berkeley: University of California Press.

———. 1988. The economic importance and marketing of forest and fallow products in the Iquitos region. Swidden-fallow agroforestry in the Peruvian Amazon. *Advances in Economic Botany* 5:74–89.

Padoch, C. and W. Jong. 1987. Traditional agroforestry practices of native and ribereño farmers in the lowland Peruvian Amazon. In H. L. Gholz, ed., *Agroforestry: Realities, Possibilities, and Potentials*, pp. 179–195. Dordrecht, The Netherlands: Martinus Nijhof.

———. 1989. Production and profit in agroforestry practices of native and ribereño farmers in the lowland Peruvian Amazon. In J. O. Browder, ed., *Fragile Lands of Latin America*, pp. 102–114. Boulder, Colo.: Westview Press.

———. 1990. Santa Rosa: The impact of the forest products trade on an Amazonian village. New directions in the study of plants and people. *Advances in Economic Botany* 8:151–158.

———. 1993. Diversity, variation, and change in ribereño agriculture. In K. H. Redford and

C. Padoch, eds., *Conservation of Neotropical Forests: Working from Traditional Resource Use*, pp. 158–174. New York: Columbia University Press.

Pahlen, A. V., W. E. Kerr, and H. Noda. 1979. Melhoramento de hortaliças na Amazônia. *Ciência e Cultura* 31 (1): 17–24.

Panero, R. 1968. Um sistema Sul-Americao de "grandes lagos." *Revista Brasileira de Política Internacional* 11 (41–42): 33–50.

Petrere, M. 1978. Pesca e esforço de pesca no estado do Amazonas. II. Locais, aparelhos de captura e estatística de desembarque. *Acta Amazonica (Suplemento)* 2 (3).

———. 1983. Relationships among catches, fishing effort, and river morphology for eight rivers in Amazonas State (Brazil), during 1976–1978. *Amazoniana* 8 (2): 281–296.

———. 1983. Yield per recruit of the tambaqui (*Colossoma macropomum* Cuvier, 1818). *Journal of Fish Biology* 22:133–144.

———. 1985. *Migraciones de Peces de Agua Dulce en America Latina*. Rome: Copescal/FAO.

———. 1986. Amazon fisheries: I. Variations in the relative abundance of tambaqui (*Colossoma macropomum* Cuvier, 1818) based on catch and effort data of the gill-net fisheries. *Amazoniana* 9:527–547.

———. 1986. Amazon fisheries: II. Variations in the relative abundance of tucunaré (*Cichla ocellaris, C. temensis*) based on catch and effort data of the trident fisheries. *Amazoniana* 10:1–13.

———. 1989. Fish stock management in the Amazon. *Annals of Amazonia: Facts, Problems, and Solutions* 1:391–401.

———. 1989. River fisheries in Brazil: A review. *Regulated Rivers: Research and Management* 4:1–16.

Pfeiffer, W. C. and L. D. Lacerda. 1988. Mercury inputs into the Amazon region, Brazil. *Environmental Technology Letters* 9:325–330.

Pfeiffer, W. C., L. D. Lacerda, and O. Malm. 1989. Mercury concentrations in inland waters of gold-mining areas in Rondônia, Brazil. *Science of the Total Environment* 87/88: 233–240.

Piedade, M. T. F., W. J. Junk, and S. P. Long. 1991. The productivity of the C4 grass *Echinochloa polystachya* on the Amazon floodplain. *Ecology* 72 (4): 1456–1463.

Piedade, M. T. F., W. J. Junk, and J. A. N. Mello. 1992. A floodplain grassland of the central Amazon. In S. P. Long, M. B. Jones, and M. J. Roberts, eds., *Primary Productivity of Grass Ecosystems of the Tropics and Subtropics*, pp. 127–158. London: Chapman and Hall.

Plotkin, M. J. 1988. The outlook for new agricultural and industrial products from the tropics. In E. O. Wilson and F. M. Peter, eds., *Biodiversity*, pp. 106–116. Washington, D.C.: National Academy Press.

Plotkin, M. J. and M. J. Balick. 1984. Medicinal uses of South American palms. *Journal of Ethnopharmacology* 10:157–179.

Prance, G. T. 1978. The origin and evolution of the Amazonian flora. *Interciencia* 3 (4): 207–222.

———. 1979. Notes on the vegetation of Amazonia: III. The terminology of Amazonian forest types subject to inundation. *Brittonia* 31:26–38.

———. 1980. A terminologia dos tipos de florestas amazônicas sujeitas a inundação. *Acta Amazonica* 10 (3): 495–504.

———, ed. 1982. *Biological Diversification in the Tropics*. New York: Columbia University Press.

Prance, G. T. and T. E. Lovejoy, eds. 1985. *Key Environments: Amazonia*. Oxford, England: Pergamon Press.

Pritchard, P. C. H. 1979. *Encyclopedia of Turtles*. Neptune City, N.J.: TFH.

Revilla, J. D. 1981. Aspectos florísticos e fitosociológicos da floresta inundável (igapó) da praia grande, Rio Negro, Amazonas, Brasil. Master's thesis. Manaus: Instituto Nacional de Pesquisas da Amazônia.

Ribeiro, M. C. L. B. and M. Petrere. 1990. Fisheries ecology and management of the jaraqui (*Semaprochilodus taeniurus, S. insignis*). *Regulated Rivers: Research and Management* 5:195–215.

Roosevelt, A. C. 1987. Chiefdoms in the Amazon and Orinoco. In R. D. Drennan and C. A. Uribe, eds., *Chiefdoms in the Americas*, pp. 153–184. Lanham, Md.: University Press of America.

Roosevelt, A. C., R. A. Houseley, and M. I. Silveira. 1991. Eighth millennium pottery from a prehistoric shell midden in the Brazilian Amazon. *Science* 254:1621–1624.

Ross, E. B. 1978. The evolution of the Amazon peasantry. *Journal of Latin American Studies* 10 (2): 193–218.

Rylands, A. B. 1993. *Marmosets and Tamarins: Systematics, Ecology, and Behaviour.* Oxford: Oxford University Press.

Santos, G. M., M. Jegu, and B. Merona. 1984. *Catálogo de Peixes Comerciais do Baixo Rio Tocantins.* Manaus: Projeto Tucuruí, Eletronorte, Instituto Nacional de Pesquisas da Amazônia.

Santos, R. 1980. *História Econômica da Amazônia (1800–1920).* São Paulo: T. A. Queiroz.

SEICOM. 1988. *Programa de Controle Ambiental da Garimpagem No Rio Tapajós, Concepção Preliminar.* Belém: Secretaria do Estado de Indústria, Comércio e Mineração.

SESPA. 1988. *Projeto Mercúrio, Relatório Parcial.* Belém: Secretaria de Saúde Pública do Estado do Pará.

Shrimpton, R. and R. Giugliano. 1979. Consumo de alimentos e alguns nutrientes em Manaus, Amazonas, 1973–1974. *Acta Amazonica* 9 (1): 117–141.

Shukla, J., C. Nobre, and P. Sellers. 1990. Amazon deforestation and climatic change. *Science* 247:1322–1325.

Sick, H. 1985. *Ornitologia Brasileira.* Brasília: Editora Universidade de Brasília.

Silva, R. 1986. Como repensar o garimpo na Amazônia? *Pará Desenvolvimento* 19:3–10.

———. 1988. *Contaminação por Mercúrio nos Garimpos Paraenses.* Belém: Departamento Nacional de Pesquisa Mineral.

Silva, R., M. Souza, and C. Bezerra. 1988. *Contaminação por Mercúrio nos Garimpos Paraenses.* Belém: Departamento Nacional de Pesquisa Mineral.

Sioli, H. 1967. Studies in Amazonian waters. *Atas do Simpósio Sobre a Biota Amazônica (Limnologia)* 3:39–50.

———., ed. 1984. *The Amazon: Limnology and Landscape Ecology of a Mighty Tropical River and Its Basin.* The Hague: Dr. W. Junk Publishers.

Smith, N. J. H. 1979. Aquatic turtles of Amazonia: An endangered resource. *Biological Conservation* 16:165–176.

———. 1980. Antrosols and the human carrying capacity in Amazonia. *Annals of the American Association of Geographers* 70 (4): 533–566.

———. 1981. *Man, Fishes, and the Amazon.* New York: Columbia University Press.

Smith, N. J. H., E. A. S. Serrão, P. T. Alvim, and I. C. Falesi. 1995. *Amazonia: Resiliency and Dynamism of the Land and Its People.* United Nations University Press, Tokyo.

Smith, N. J. H., J. T. Williams, D. L. Plucknett, and J. Talbot. 1992. *Tropical Forests and Their Crops.* Ithaca, N.Y.: Cornell University Press.

Spix, J. B. and C. F. P. Martius. 1938. *Viagem Pelo Brasil.* 3 vols. Rio de Janeiro: Imprensa Nacional.

Spruce, R. 1908. *Notes of a Botanist on the Amazon and Andes.* London: Macmillan.

Sternberg, H. O'R. 1956. *A Água e Homem na Várzea do Careiro.* Rio de Janeiro: Universidade do Brasil.

Terborgh, J. 1983. *Five New World Primates: A Study in Comparative Ecology.* Princeton, N.J.: Princeton University Press.

Terborgh, J., S. K. Robinson, and T. A. Parker. 1990. Structure and organization of an Amazonian bird community. *Ecological Monographs* 60 (2): 213–338.

Vari, R. P. 1988. The Curimatidae, a lowland neotropical fish family (Pisces: Characiformes): Distribution, endemism, and phylogenetic biogeography. In P. E. Vanzolini and R. R. Heyer, eds., *Proceedings of a Workshop on Neotropical Distribution Patterns*, pp. 343–377. Rio de Janeiro: Academia Brasileira de Ciências.

Wallace, A. R. 1853. *Narrative of Travels on the Amazon and Rio Negro.* London: Reeve and Company.

Worbes, M., H. Klinge, J. D. Revilla, and C. Martius. 1992. On the dynamics, floristic subdivision, and geographical distribution of várzea forests in central Amazonia. *Journal of Vegetation Science* 3:553–564.

Zimmerman, B. L. and M. T. Rodrigues. 1990. Frogs, snakes, and lizards of the INPA-WWF Reserve near Manaus, Brazil. In A. H. Gentry, ed., *Four Neotropical Rainforests*, pp. 426–454. New Haven, Conn.: Yale University Press.

PHOTO CREDITS

CHAPTER 1

The flooded forest. Michael Goulding
Most endangered habitat. Michael Goulding
Deforestation. Nigel Smith
Rio Madeira. Michael Goulding
Rio Tapajós. Michael Goulding
Rio Negro. Michael Goulding
Uplands meet the river. Nigel Smith
Canoeing the canopy. Michael Goulding
Showcase of the floating meadows. Michael Goulding
Floodplain lake. Nigel Smith
Flooded out. Michael Goulding
Symbol of the flooded forest. Michael Goulding

CHAPTER 2

Pictographs. Nigel Smith
What the first Europeans saw. Michael Goulding
Harvesting maize. Nigel Smith
Medicinal plants. Michael Goulding
Turtle eggs. Michael Goulding
Rubber tapper. Michael Goulding

CHAPTER 3

Jute. Nigel Smith
Manaus. Michael Goulding
Log rafts. Michael Goulding
The refuse of deforestation. Michael Goulding
Drowned forests. Michael Goulding
Gold fever. Michael Goulding

CHAPTER 4

Victims of deforestation (macaws). Michael Goulding
The strangest bird (hoatzin). Michael Goulding
Wading birds. Michael Goulding
Target of the cage-bird trade (troupial). Michael Goulding
World's smallest monkey (pygmy marmoset). Michael Goulding
Endemic primate (uacari). Michael Goulding
World's largest rodent (capybara). Michael Goulding
Teiid lizard. Michael Goulding
Once populous (black caiman). Michael Goulding
Seasonal users (side-necked turtles). Michael Goulding
Arthropod heaven. Michael Goulding

CHAPTER 5

Catfish migration. Michael Goulding
An extraordinary migrator (dourada catfish). Michael Goulding
Voracious predator (bôto dolphin). Michael Goulding
Favored fish habitat. Michael Goulding
Fish food. Michael Goulding

CHAPTER 6

Local versus export fisheries. Michael Goulding
Bélem. Nigel Smith
Fishing for rock-bacu. Michael Goulding
Fishing with gillnets. Michael Goulding
Fishing with castnets. Michael Goulding
Tambaqui. Michael Goulding
Pirarucu. Michael Goulding
Fish catches in Manaus. Michael Goulding
Catfish at Teotônio rapids. Michael Goulding
Working the aquarium trade. Michael Goulding
Target of aquarium trade (cardinal neon). Michael Goulding
Tucunaré. Michael Goulding

CHAPTER 7

Palms, the "trees of life." Michael Goulding
Assaí palm and fruit. Michael Goulding
Nuts from the floodplain (sapucaia). Nigel Smith; Michael Goulding
Uxi fruits. Nigel Smith
Fish baits. Michael Goulding
Fuelwood. Nigel Smith

CHAPTER 8

Small livestock. Michael Goulding
Floating meadows as pasture. Michael Goulding
When the forage fails. Nigel Smith
When the floods come. Nigel Smith; Michael Goulding
Water buffalo. Michael Goulding
Best soils. Michael Goulding
Coping with floods. Michael Goulding; Nigel Smith

Preparing manioc flour. Michael Goulding
Grove of bananas. Michael Goulding

CHAPTER 9

Primary challenge to conservation. Michael Goulding
Forests and the fisheries. Michael Goulding
Fish, source of animal protein. Michael Goulding
Confusions over land title. Nigel Smith
Ominous shadows. Michael Goulding
What fate? Michael Goulding

INDEX

ABOUT THE AUTHORS

Michael Goulding (Ph.D., UCLA, 1978) is senior scientist at the Rainforest Alliance, a conservation and research NGO based in New York City, and director of its Amazon Rivers Program. He formerly worked for Brazil's two leading Amazonian research institutions, the National Institute of Amazonian Research (INPA) and the Goeldi Museum. He has also been extensively involved in natural history films about the Amazon. Publications include *The Fishes and the Forest* and *Amazon: The Flooded Forest.*

Nigel J. H. Smith (Ph.D., Berkeley, 1976) is Professor of Geography at the University of Florida, Gainesville. He was previously a researcher at the National Institute of Amazonian Research, Worldwatch Institute, and the World Bank. Publications include *Man, Fishes, and the Amazon* and *Rainforest Corridors.*

Dennis J. Mahar (Ph.D., University of Florida, Gainesville, 1970) is Resident Representative in Brazil for the World Bank. He was previously Chief of the World Bank's Environmental Division for Latin America and the Caribbean. Publications include *Government Policies and Deforestation in Brazil* and *Frontier Development in Brazil: A Study of Amazonia.*